vered Islands

The Un-Discovered Islands

An Archipelago of Myths and
Mysteries, Phantoms and Fakes

Malachy Tallack

Illustrated by Katie Scott

Picador / New York

picadorusa.com • picadorbookroom.tumblr.com
twitter.com/picadorusa • facebook.com/picadorusa

Picador® is a U.S. registered trademark and is used by Macmillan Publishing Group,
LLC, under license from Pan Books Limited.

For book club information, please visit facebook.com/picadorbookclub
or email marketing@picadorusa.com.

Designed by Jamie Keenan

Endpaper art illustrated by Katie Scott
Endpaper designed by Jamie Keenan

The Library of Congress Cataloging-in-Publication Data is available upon request.

ISBN 978-1-250-14844-5 (paper over board)
ISBN 978-1-250-14845-2 (ebook)

Our books may be purchased in bulk for promotional, educational, or business use. Please contact your
local bookseller or the Macmillan Corporate and Premium Sales Department at 1-800-221-7945,
extension 5442, or by email at MacmillanSpecialMarkets@macmillan.com.

Originally published in the United Kingdom by Polygon, an imprint of Birlinn Ltd.

First U.S. Edition: November 2017

10   9   8   7   6   5   4   3   2   1

Contents

Introduction

I remember well the motto of the Anderson High School in Lerwick, displayed on the brightly coloured crest that was fixed to the gates outside. 'Dö weel and persevere', it counselled. At some point we pupils must have been told the origin of these words, for they were intimately tied to the place itself. 'Dö weel and persevere' was the formative advice given in 1808 to the young man Arthur Anderson, later to be the industrialist Arthur Anderson, co-founder of the P&O shipping company, member of parliament for Orkney and Shetland, and benefactor of the school that still bears his name.

It was not a particularly stirring piece of advice. To me it sounded half-hearted, like the words of an inattentive father patting his son absent-mindedly on the head. But the story of Anderson's rise from poverty to philanthropy was supposed to inspire young Shetlanders. It was part of the history of our school and the history of our islands. The implication was that, if heeded, these words could help shape our futures too. Hard work and perseverance: those were the lessons that would lead us forward.

Accompanying that motto on the crest were three Viking images – an axe, a longship and a flaming brand – alongside another, more ambiguous inscription. On a yellow scroll across the centre of the emblem were three words in Latin that pointed to a rather different part of our history. '*Dispecta est Thule*': Thule was seen.

Though I passed through those gates countless times in my years at school, no teacher ever explained the Latin words they bore, and I never bothered to ask. From somewhere, I had gathered a vague notion that Thule was supposed to be the edge of the world, and that somehow Shetland was it, or at least it once had been. But in my youthful head that word was connected most closely with the Thule Bar down at the harbour, a far more mysterious and tantalising place for a teenage boy.

It was not until several years later, when school was long behind me, that I learned the origin of this motto. Thule was indeed the edge of the world, but it was more than that. It was an island once believed to be real but now absent from the maps. It was a place that was no longer a place. The words themselves came from the Roman historian Tacitus, whose father-in-law, Agricola, was governor of Britain in the late first century AD. Sailing north of mainland Scotland, Agricola had seen Shetland on the horizon and believed it to be Thule, the northernmost point in the classical world. He pinned the label to the islands, but it didn't stick for long. Thule was seen, and then once again it disappeared.

In hindsight it seems odd that such a phrase was considered a suitable decoration for those gates, since its message so obviously clashed with the one that accompanied it on the crest. According to the school motto, Shetland was as pivotal a place as we, its sons and daughters, wished to make it. Arthur Anderson was an important man – his ships had sailed the world's oceans – and like him we could go anywhere and do anything. But in those three words from Tacitus, Shetland lost its identity altogether and tumbled off the edge of the map. Hitched to the idea of Thule, we barely even existed at all. It was a peculiar contradiction, but something in that unreal geography appealed to me.

Later I found that the oceans are full of such places: islands discovered and then un-discovered. They have existed in every part of the world, and some appeared on maps for many centuries before finally being erased. These islands have not been lost to rising seas or to earthquakes; they are not the victims of natural disasters. These islands are human in origin, the products of imagination and error.

Gathered in this book is a whole archipelago of un-discovered islands, grouped into six sections. The first are *Islands of Life and Death*: mythical places, confined to stories. *Setting Out* introduces islands found by early travellers in the Atlantic and Pacific, when few people knew the world beyond their own shores. The third group emerged during the *Age of Exploration*, as European sailors began to crisscross the globe with increasing regularity. The fourth are *Sunken Lands*, once thought to have been submerged; while the fifth are *Fraudulent Islands*, invented by hoaxers and liars. The sixth and final group are *Recent Un-Discoveries*, made during the twentieth and twenty-first centuries.

Each of these places has its own story. None is exactly alike. Some have helped to shape entire cultures, while others have been barely noticed. Some are strange and fabulous, while others are utterly believable. All of them reflect in some way the values of their age, and all of them have enriched the geography of the mind. This book seeks to celebrate and commemorate these un-discovered islands, and, through them, tell the story of how we have created our image of the world.

Isla

of Lif

De

Islands of Life and Death

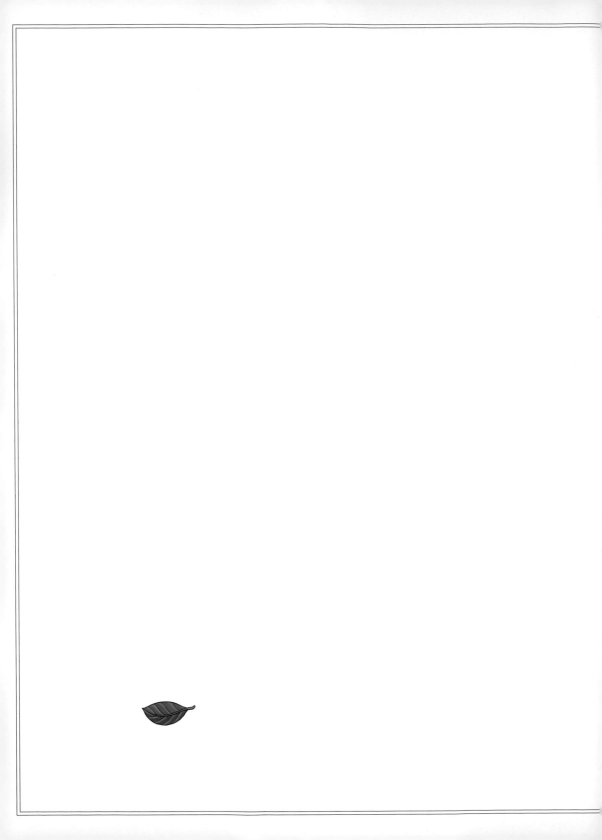

Islands of Life and Death

FACED WITH THE SKY we imagine gods; faced with the ocean we imagine islands. Absence is terrifying, and so we fill the gaps in our knowledge with invented things. These bring us comfort, but they conflict, too, with our desire for certainty and understanding. And sometimes that desire gives us back the absences we sought to fill.

For as long as people have been making stories, they have been inventing islands. In literature and in legend, they are there from the very start. For societies living at the sea's edge, the dream of other shores is the most natural dream there is. Polynesians, Marsh Arabs, the ancient Greeks, the Celts: all imagined lands beyond their horizon. All of them told stories of islands.

These places were not quite like the everyday world. They were supernatural regions, where the lines between life and death were blurred. The ocean divides us from other lands, just as death divides us from the living. The crossing can be made, but only once. Islands, then, are perfect metaphors for other worlds and afterlives. They are separate and yet connected; they are distant and yet tangible. The sea of death is cluttered with imaginary islands.

Today, we try to draw strict lines between facts and fictions. But myth, superstition and religion have been part of human life for as long as we have been human. They have shaped our thinking and guided our actions. The way we comprehend our existence is indivisible from the stories we have told ourselves. So while the islands in this chapter may be mythical, they were no less real for that.

THE NOTION OF A paradise on Earth has long been part of European mythical traditions, and in Homer's *Odyssey* we find one of the oldest extant versions of the story. There, Elysium, or the Elysian Plain, is the land to which those favoured by the gods are brought. According to Proteus, the Old Man of the Sea, people there 'lead an easier life than anywhere else in the world, for in Elysium there falls not rain, nor hail, nor snow, but Oceanus breathes

Plato, in the fourth century BC, Elysium was most commonly imagined as an island or archipelago in the western ocean. It was known as the White Isle, or the Isles of the Blessed, and some considered it a place to which all could aspire.

In Plato's dialogue *Gorgias*, Socrates outlines his own belief, in terms that clearly anticipate the Christian religion yet to be born. After death, he says, body and soul become

The Isles of the

ever with a west wind that sings softly from the sea, and gives fresh life to all men'. This, then, was not a place beyond death, but an alternative to it.

The ancient Greeks did not have one single version of this story, however. It was an evolving and multifarious idea. By the time of

separated, but each retains the character it had when alive. The fat remain fat; the scarred remain scarred. At least for a time. Equally, 'when a man is stripped of the body, all the natural or acquired affections of the soul are laid open to view'. Unlike the body, however, the soul must face judgement after death, a task undertaken by three sons of Zeus. Aeacus judged those from the west and Rhadamanthus those from the east, with Minos as the final arbiter. Anyone who has 'lived unjustly and impiously shall go to the house of vengeance and punishment, which is called Tartarus'; whereas, 'he who has lived all his life in justice and holiness shall go, when he is dead, to the Islands of the Blessed, and dwell there in perfect happiness out of the reach of evil'.

The fact is, Socrates told them, 'that to do injustice is more to be avoided than to suffer injustice, and . . . the reality and not the appearance of virtue is to be followed above all things, as well in public as in private life'. Only then can one guarantee a passage to paradise.

The Celts too believed in a blessed island, according to the earliest recorded stories. In fact, there were several such islands, including Tír na nÓg, the land of eternal youth.

Socrates knew that his listeners – the rhetoricians Gorgias, Callicles and Polus – considered this story to be a myth. But he suggested they reconsider. His own life had been well lived, he claimed, and he felt ready to present his soul 'whole and undefiled before the judge'. Did they share that confidence in themselves?

It was there to which the young warrior poet Oisín eloped with Niamh, the daughter of a sea god called Manannán mac Lir. On returning to Connemara to visit his family, three years after the marriage, Oisín discovered that a year in Tír na nÓg was the same as a century in Ireland. His family were long dead.

Other such realms were often inter-changeably. There was the island of Mag Mell, akin to Homer's Elysium, where deities and favoured mortals lived without pain or sickness. There was, too, Emhain Ablach and its Welsh equivalent Ynys Afallon, the island of apples. Fruitfulness, for the Celts, was a key feature of the place.

In medieval times, that island of apples became known most famously as Avalon. It was there that King Arthur's sword Excalibur was forged, and it was there where the king himself would later retire after being wounded at the Battle of Camlann. Just as for the early Greeks, the heroic Arthur had earned his place on the blessed isle, and his journey to it was an alternative to death. According to legend, the king would one day return from Avalon to fight for his people: a kind of Celtic messiah.

It is from the twelfth-century cleric Geoffrey of Monmouth that much of the story of Arthur is derived. In his *Vita Merlini*, Geoffrey described Avalon in some detail – detail that has been drawn directly from the Roman tradition of the Fortunate Isles and the Greek traditions of Elysium, the garden of Hesperides and the Isles of the Blessed.

The Island of Apples gets its name
'The Fortunate Island' from the fact
that it produces all manner of plants
spontaneously. It needs no farmers
to plough the fields. There is no
cultivation of the land at all beyond
that which is Nature's work. It pro-
duces crops in abundance and grapes
without help; and apple trees spring
up from the short grass in its woods.
All plants, not merely grass alone,
grow spontaneously; and men live a
hundred years or more.

In cartography, the Fortunate Isles became associated with the Canaries, and medieval maps often rendered that archipelago as *Insula Fortunata*. But the mythical origins of the name were not forgotten. Although Christian teaching insisted that paradise lay in a super-natural realm, the idea of a promised land on Earth never left the European imagination. The fruitful isle remained on the western horizon. In England, the blissful land of Cockaigne was the subject of countless stories and poems; in Germany it was Schlaraffenland, the land of milk and honey; and in Spain it was Jauja, a name now attached to a small city in Peru.

As European explorers began pushing further into the Atlantic in the fourteenth and fifteenth centuries, many expected to find such an idyll somewhere out there. Later, after Columbus, that expectation seemed for a time to have been met, and the language and imagery once associated with the Isles of the Blessed were bestowed upon the newly discovered continent. The promised land had been found, it seemed, and it was called America.

14

AFTER DEATH, THE bodies of islanders from Mabuiag in the Torres Strait would be taken outside and laid on a platform. Clan members of the dead person's spouse would then watch over them, to ensure that the spirit, or *mari*, had properly evacuated the corpse. They would also protect it from the hungry mouths of lizards.

After five or six days, the body, which by then would be putrid, was decapitated. The head would be placed in a nest of termites, or in water, to remove the flesh. The rest of the corpse remained on the platform, covered in grass, until only the bones were left.

Once cleaned, the skull would be coloured red and placed in a basket, decorated with feathers and hair. The deceased's in-laws, who were in charge of these rituals, would then perform an elaborate ceremony in front of the dead person's family. For this they would paint themselves black and cover their heads with leaves, before presenting the skull to the closest relative. A chant would be offered to console the mourners:

When the wind comes from the north the sky is black with clouds and there is much wind and pouring rain, but it does not last long, the clouds blow over and there is fine weather once more.

Ki

Other islands of the western Torres Strait had rituals that differed slightly from this one. In some, the body would be buried in a shallow grave, or else desiccated and mummified, while on others the skull would

----

be adorned with beeswax and shells. On one island – Muralug – a widow was expected to carry the skull of her husband in a bag for a year after his death, while other family members might wear his bones as ornaments, or keep them safe in their houses.

mari would be carried there on the prevailing south-easterly winds.

Upon arrival, the spirit was met by the ghost of an acquaintance – usually their most recently deceased friend – who would take them into hiding until the next new moon. At that time they would emerge and be introduced to the other spirits of the island, who would each hit them upon the head with a stone club This seemingly un-welcoming act was, in essence, an initiation ceremony, and from that moment on the *mari* was a *markai*: a ghost proper.

Some believed the *markai* spent their time in treetops, cry-ing, perhaps in the form of flying foxes. But most agreed that the afterlife was not so different from this one, and that the spirits remained in human form. During the day they would hunt for fish with spears, and in the early evening they might dance on the

One element was common to all, however: the belief in an island of ghosts, to which the dead person's spirit would travel. That island, called Kibu, was beyond the northwest hori-zon, and once it had escaped from the body the

beach. The *markai* could also catch turtles and dugong (a marine mammal related to the manatee) by creating waterspouts, up which the animals would be drawn.

But ghosts were not restricted to Kibu. They could return home temporarily if they wished, and sometimes they would even go to war with the living. Islanders often invoked the *markai*, whether individually, through divination and spirit consultation, or in ceremonies such as the 'death dance', which was usually held several months after a person had passed away.

In Mabuiag, these ceremonies were called the *tai*, or simply the *markai*, and were held on the nearby uninhabited island of Pulu. Often they would mark the deaths of several people at once, and the details of the performance would depend on who and how many were being commemorated. The essence of the ceremony, however, was the representation of the dead by the living. Those taking part would rub their bodies in charcoal and décorate themselves with leaves and feathered headdresses, until they were fully disguised. Each would take on the character of a specific person, and would become, in the minds of the audience, that person's ghost.

The performers carried bows and arrows, or brooms, and danced and jumped before the spectators. There was an odd, slapstick element to these dances, with one performer skipping and falling over, while others loudly

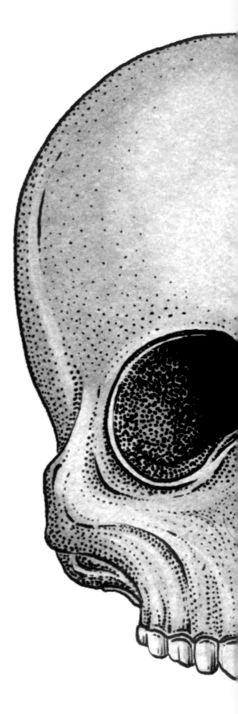

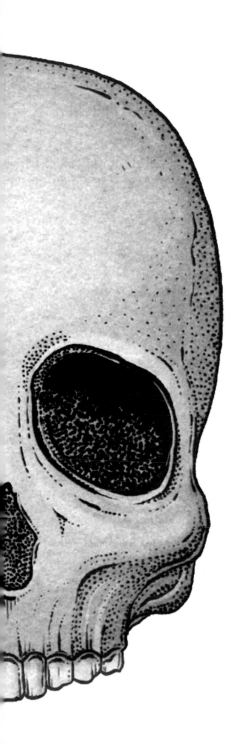

broke wind. The ceremony concluded with the beating of drums and with a great feast.

Throughout the *tai*, the performers were imitating and personifying the dead. It was a form of consolation for the relatives, and an insistence on the continuation of that person's spirit. It was believed that the ghost was present within the dancers, and that it would continue to be part of the world. This connection was crucial. The divide between life and afterlife was like that between islands: it was real, but not insurmountable. It could be crossed. Like Kibu itself, the ghost world was accessible and comprehensible. But that accessibility would not last forever.

The rites and beliefs of the Torres Strait islanders were recorded by members of the Cambridge Anthropological Expedition at the very end of the nineteenth century. But already then things were changing rapidly. The islands' government, together with missionaries, were eager to suppress and replace native customs. More spiritually and physically hygienic forms of burial were insisted upon, and the traditional beliefs were gradually replaced by Christian ones.

Kibu too was replaced, of course, by a heaven that was entirely unlike the islanders' own world. The afterlife today lies not just over the north-west horizon but skyward, detached entirely from the islands and from the sea. Unlike Kibu, heaven is unimaginable, and the ghosts of the dead are now gone for good.

Hawaiki

WHEN MĀORI PEOPLE first began to communicate with Europeans in the eighteenth century, they insisted that New Zealand was not their original home. Instead, they explained, their ancestors had come from Hawaiki, an island somewhere over the northeast horizon. What's more, they had not arrived in the distant past, but only a few hundred years previously.

The details of this migration were not entirely clear. Different tribal groups, or *iwi*, told different versions of the story. And though their cultural memory was rich in detail, many Māori were understandably reluctant to share such important knowledge with settlers, especially since those settlers were also demanding to share their land.

In the best known version of the country's early history, a fisherman and explorer called Kupe discovered New Zealand more than a thousand years ago. He arrived there by accident, while chasing a giant octopus south across the ocean. Kupe then returned to Hawaiki and told his people about this new land in the south, which he called Aotearoa, the 'long white cloud'. Around **1350**, following the instructions he had given, a 'great fleet' of seven large canoes set out to make the crossing back to Aotearoa. The passengers in those canoes were the ancestors of today's Māori.

The problem with this story is that it wasn't reliable. It was a constructed history, an amalgamation of many different tales pieced together by an ethnologist, Stephenson Percy Smith, in the late nineteenth and early twentieth century. Smith was a thorough researcher, but his conclusions were misleading. Rather than accept the inconsistencies and mythical elements that he found within traditional stories, Smith tried instead to iron them out and present the results as fact. In doing so, he built a narrative that was neither historically accurate nor truly representative of what the Māori themselves actually believed.

As it turns out, the date Smith proposed for the 'great fleet' was not that far wrong. According to the most recent evidence, the first people to settle in the country arrived around **1280**, though not in a single flotilla but more likely in several groups, perhaps over a period of decades. The Polynesians were highly skilled navigators, and there could have been contact for some time between the new and old home. In total, there may have been as few as **200** people among those first immigrants.

The Māori's geographical origin can also

now be pinpointed with a fair degree of certainty. They came from eastern Polynesia: specifically, the Cook and Society Islands. Which might provide a simple answer to the question of Hawaiki. Except that it doesn't. For Hawaiki is not simple at all. In traditional stories it is a multifaceted idea that cannot be pinned down to a single location. This island was not just the migrants' point of departure, it was part of their luggage – that rich, mythical tradition with which they arrived.

In 1793, a Māori chief called Tuki Tahua was asked to draw a map of New Zealand for the governor of New South Wales and Norfolk Island. This he did, with an impressive degree of accuracy. But in addition to the physical features of the land, Tuki also included what he called a 'spirits road', which traced the line of mountain ranges from the far south right up to the North Cape. This was the path that one would follow after death, he explained, which led ultimately to Te Reinga Wairua, 'the leaping place of the spirits'. From that final point of land, at the tip of the North Island, each spirit would dive into the ocean then swim towards the underworld, where they would find Hawaiki.

But this island unfolds still further, for these were the words with which newborn babies were traditionally welcomed into the world:

> *E taku pōtiki, kua puta mai rā koe i te toi*
> *i Hawaiki.*
> *My child, you are born from the source,*
> *which is at Hawaiki.*

Both afterlife and prelife, Hawaiki surrounds the Māori. It is the the place from which the spirit comes and to which it returns; it is the source and the destination. In some stories, it is also the place in which the very first human was created: a kind of Eden, where gods still dwell. The precise way in which the island is portrayed – the balance between physical homeland, spiritual origin and underworld – varies greatly, depending on the story being told and the local culture of the teller. But Hawaiki is a shared idea; it ties people together. And not just within New Zealand.

Eastern Polynesia is among the most recently inhabited parts of the world. Many of the islands of that region which stretches from Hawaii in the north to Easter Island in the east and New Zealand in the south – were populated only within the past fifteen hundred years or so. The traditions of these places are closely related and interlinked, and the notion of an origin elsewhere has remained fresh in the thinking of their people. Where most cultures have myths of creation, the Polynesians have myths of migration.

For the Māori, Hawaiki is a place of goodness. It is the place from which their people, their traditions and their culture derive. It is both real and imagined, both geographical and mythical. Yet it does not divide them from their current home, for it is within as well as without. It connects, in time and in place.

A T THE CONFLUENCE OF two great rivers, the Tigris and Euphrates, there was once a wetland that covered thousands of square miles, the largest of its kind in all of Western Eurasia. This region – once part of Mesopotamia, now southern Iraq – was the birthplace of modern civilisation, and home to the Ma'dān people, known as the Marsh Arabs. The Ma'dān are descended from Babylonian, Sumerian and Bedouin cultures, and for five millennia their lifestyles barely changed. The way in which they lived was defined, always, by the place in which they lived.

That place was one of shallow lagoons, gangling bulrushes and low floodplains. It was a strange world, where buffalo waded among the reeds and pelicans flocked overhead; beneath the waves swam deadly serpents. Few outsiders ever knew this place, and until the mid-twentieth century it was a mysterious, half-mythical location. Gavin Maxwell travelled to the marshes in 1956, from where he brought back Mijbil, the otter that was to be at the centre of his most famous book, *Ring of Bright Water*. That otter was of a subspecies previously unknown to science, and today it carries his name: *Lutrogale perspicillata maxwelli*.

Maxwell wrote of his time in the marshes in *A Reed Shaken by the Wind*, and in his first encounters with the place he seemed confused about how to respond. It was both repellent and beautiful to him, like nowhere he had ever seen before. 'It was in some ways a terrible landscape',

he wrote, 'utterly without human sympathy, more desolate and inimical than the sea itself'. And yet, two days later, that 'terrible landscape' had become:

> *a wonderland, and the colours had the brilliance and clarity of fine enamel. Here in the shelter of the lagoons the reeds, golden as farmyard straw in the sunshine, towered out of water that was beetle-wing blue in the lee of the islands or ruffled where the wind found passage between them to the full deep green of an uncut emerald.*

The people of this region lived with what their home provided. The water buffalo were not eaten, but their milk was drunk and their dung used as fuel and as cement. The Ma'dān fished with spears, kept birds to eat, and in the saturated earth they cultivated rice. The reeds that grew there in abundance were tall and strong enough to be used for making boats and building houses. From inside, the great halls they constructed – *mudhifs* – looked like the hollowed interior of a whale. Tall, curved ribs of woven reeds supported a thick, thatched skin. Everything that was needed came from the water.

The Ma'dān were Shia Muslims, and some were descendants of the prophet Mohammed himself. They believed in supernatural spirits, or *jinn*, that could take the form of snakes and other creatures. But they also retained elements

----

Hufaidh

of pre-Islamic beliefs, including stories about a magical island somewhere out in the marshes. The explorer Wilfred Thesiger visited the Ma'dān several times in the 1950s, living with them for many months at a time. On one of his stays, Thesiger was asked if he had ever heard of Hufaidh. He had, he said, but he wanted to know more. Waving towards the south-west horizon, his host told him: 'Hufaidh is an island somewhere over there. On it are palaces, and palm trees and gardens of pomegranates, and the buffaloes are bigger than ours. But no-one knows exactly where it is.'

'Has no-one seen it?' Thesiger asked. 'They have, but anyone who sees Hufaidh is bewitched, and afterwards no-one can understand his words. By Abbas, I swear it is true. One of the Fartus saw it, years ago, when I was a child. He was looking for a buffalo and when he came back his speech was all muddled up, and we knew he had seen Hufaidh.'

The Ma'dān explained that anyone searching for the island would fail to find it. The *jinn* could make it disappear at will. But Hufaidh was real, they said. The sheiks knew of it, the government knew of it; there was no room for doubt. Like many such islands, Hufaidh existed in a region bridged between life and death. It was part paradise and part hell, both of this world and of another.

But Hufaidh is no longer of this world, for the marshes are a very different place today. The draining of the wetlands began around the time of Thesiger and Maxwell's visits. Initially it was on a small scale, to increase the availability of agricultural land. But as the decades passed, more irrigation channels diverted water away from the rivers, and the marshes began to shrink. It was not until the 1990s, though, that the damage was truly and deliberately done.

Saddam Hussein hated the Ma'dān. As Shias, they were hostile to the Sunnis who were in power, and had sheltered dissidents and re-

bels. So when the first Gulf War ended, Saddam took terrible revenge. He diverted the flow of the Tigris and built a new canal to ensure the water would go elsewhere. The plan succeeded. Within two years, two-thirds of the wetlands had dried up, and by the end of the decade ninety per cent of the marshland was gone. It was an act of devastating barbarism, a human and ecological tragedy.

Thousands of miles of southern Iraq, once home to fish, plants, birds and mammals, turned to desert. A unique ecosystem was lost. And the people who depended on that ecosystem – who were, in fact, part of it – were forced to flee. In the 1950s, there were half a million Ma'dān in the region. Today, there may be only ten per cent of that number, and perhaps fewer than 2,000 living as they did for five millennia, in reed huts on the water.

After the second Gulf War, Saddam's work was undone. The embankments were destroyed, and water was allowed to flow into the marshes once more. In the years since then they have grown, slowly, and they continue to grow. Some of the species that once inhabited the region have returned, though some are extinct and can never come back. The restoration of such a place is not a simple task, and some damage cannot be undone. The culture of the Ma'dān may not be lost forever; those who remained may stay, and some who left may return. But Hufaidh – that island of palm trees and pomegranates – has gone. It has turned to sand, and scattered in the wind.

Set

Q

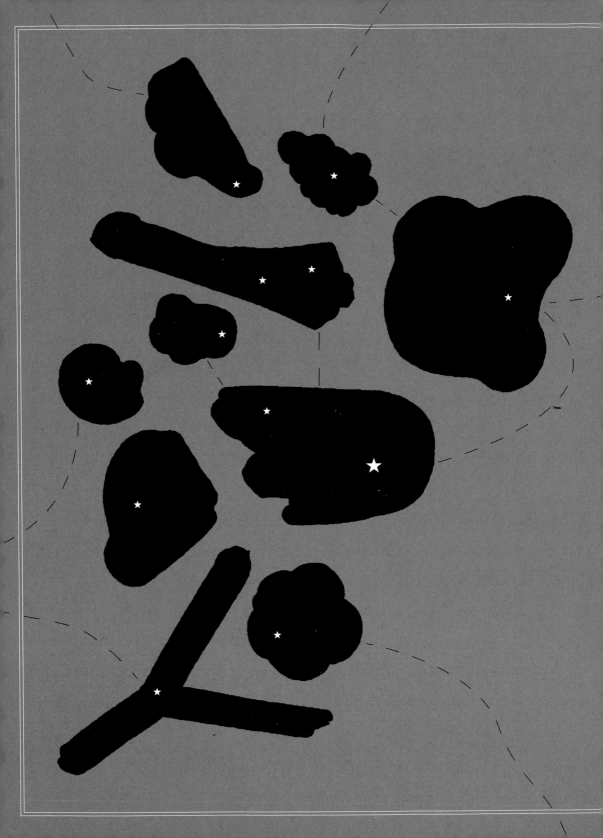

×

Setting Out

Setting Out

FOR TRAVELLERS IN THE last centuries BC and the first millennium AD, the boundaries of geographical knowledge were narrow. People understood that the world was big and that their part of it was small, but they knew little of what lay beyond. The map was hardly more than a sketch, its edges crowded with speculation. Those who did make journeys towards those edges would encounter things they had never seen or even heard about before. The ocean was a terrifying, wonderful place, where legends and facts would mingle, and where anything imaginable might be possible.

During these centuries, extraordinary journeys were taking place all over the world. In the Pacific, the Polynesians were navigating across thousands of miles, using skills their descendants still employ today. In the North Atlantic, the Norsemen were island-hopping, from Shetland to Faroe to Iceland to Greenland, and even to North America. They too developed a rare competence at sea, which took them to places no European had ever been before.

Everywhere, human beings were crossing the oceans in search of new land. Some of these journeys were recorded in writing, some in oral traditions, and others on maps. But myth and geography are difficult to prise apart after so long. Facts are hard to separate from fiction. Legendary islands appeared on charts of the Atlantic as late as the nineteenth century without any proof of their existence, yet stories of Viking expeditions to 'Vinland' more than one thousand years ago were widely considered to be false until archaeological evidence of Norse settlement was uncovered in L'Anse aux Meadows, Newfoundland, in 1960.

Some of the islands in this chapter may likewise be real places, but it is impossible now to know. Their stories are so distant, and so infused with the imaginary, that they exist today only in name. With nowhere left to go, they are true ex-isles.

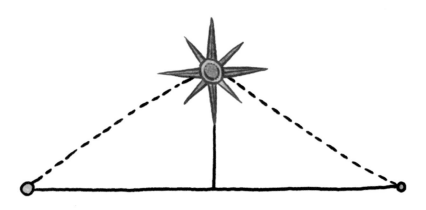

B RITAIN IS A LONG way from the
Mediterranean, and for the ancient Greeks
it was a dark and potentially dangerous land, at
the edge of the human world. But in the fourth
century BC, an explorer from the Greek colony
of Massalia – today's Marseille – claimed not
just to have reached Britain but to have gone
beyond, to the previously unknown island of
Thule.

That explorer was Pytheas, and his book
recounting the journey, *On the Ocean*, though
since lost, was widely read and remarked upon
by other classical writers. From what has been
pieced together, it seems Pytheas set out on
his travels around 330BC. He first reached the
tin-producing regions of south-west Britain and
then went onwards, taking measurements of the

sun's height along the way. When he reached
the edge of the mainland he did not turn back.
Instead, the Greek claimed to have continued,
travelling six days north to the 'farthest of all
lands', Thule. This was truly an astonishing
journey.

Among early commentators, however,
Pytheas' voyage was not looked upon with
unqualified admiration. Some expressed
considerable scepticism about the authenticity
of his reports, and in particular serious doubts
were raised about the existence of Thule.
In his *Geography* of 30AD, Strabo, another
Greek historian, was voracious in his attacks.
Repeatedly he questioned Pytheas' claims, and
described his fellow Greek as an 'arch-falsifier'.
Earlier still, in the second century BC, Polybius

wrote that Pytheas 'misled many readers' with his stories, and that 'Even Erastothenes doubted' elements of *On the Ocean*.

The suspicion that came to surround Thule was understandable, for there was much to

Thule

raise scholarly eyebrows about in Pytheas' tale of distant lands. For one thing, he claimed to be travelling in a place that many believed to be too far north for human habitation. According to Strabo, the island of Ireland 'is such a wretched place to live in on account of the cold that the regions beyond are regarded as uninhabitable'. So Thule, a place six days' sail north of Britain, was a highly implausible idea.

But later writers have been less cynical. Pytheas has been given the benefit of the doubt by many, and celebrated as a true northern pioneer. The descriptions that survive of Thule seem just about plausible enough for it to be considered a real place. But where? The evidence is thoroughly ambiguous. A six-day journey from Britain could take you to any number

of places, depending on where you left from, the direction you travelled, whether overnight stops were made, the type of craft that was used, the weather, and a multitude of other factors. Shetland, Norway, Faroe, even the Baltic, if your idea of north was somewhat confused: all could be reached within six days.

Beyond this, though, there are several other elements of the story that have fed speculation. There is, first of all, the rather fantastical description that Polybius gives, seemingly quoting directly from *On the Ocean*. In the vicinity of Thule, he writes:

> there is neither unmixed land or sea or air, but a kind of compound of all three (like the jelly-fish or Pulmo Marinus), in which earth and sea and everything else are held in suspense, and which forms a kind of connecting link to the whole, through which one can neither walk nor sail.

This is a very strange image indeed, but has most often been taken to refer to fog combined with slushy sea ice, rising and falling like the body of a jellyfish.

Another intriguing part of Pytheas' tale is the matter of daylight. According to Pliny the Elder, writing in the first century AD, in Thule 'there is no night at the summer solstice, when the sun is passing through the sign of Cancer, while on the other hand at the winter solstice there is no day. Some writers are of the opinion that this state of things lasts for six whole months together.' Discounting the latter possibility, since it could be true only at the North Pole, the suggestion here is that Thule lies somewhere in the vicinity of the Arctic Circle, which would certainly shorten the list of possibilities. Unless, of course, the description is not entirely literal. In many parts of the north, including Shetland, the skies remain pale throughout midsummer nights, while in winter the sun seems barely to lift the darkness at all. A southern visitor could quite easily describe these periods in the language used by Pliny.

Over the centuries, the most popular candidates for the title of Thule have been Shetland, Norway and Iceland, and for a long time it was the last possibility that was favoured. On maps, the two names were for a long time interchangeable. The Venerable Bede, in the early eighth century, was certain of the correlation, as was Christopher Columbus in the fifteenth. Gerardus Mercator, on his world map of 1569, evidently considered Iceland and Thule to be one and the same; and Barry Cunliffe, formerly professor of European archaeology at Oxford, has recently proposed the very same theory.

For poets and explorers, cartographers and historians, Iceland has seemed to fulfil the most important requirements of a Thule: namely, that it is far away and strange. But in accepting this they have chosen to ignore two factors. First, that reaching Iceland from Britain in six days, more than 2,000 years ago, would have been extraordinarily difficult. And second, that according to Strabo and others, Thule was populated by farmers who brewed a drink from honey; but at the time Pytheas travelled, Iceland was uninhabited. It had no people and no bees.

The problem, for those who have sought to pin Thule down, is that nowhere is quite right. The few clues available add up only to contradictions, and to a vague, uncertain shape. To try and unravel these contradictions or to add detail to the image is to fail almost at once. Pytheas may have travelled to Shetland or to Faroe; he may have reached Norway, Iceland or the Baltic; he may even have constructed his story entirely out of rumours and fantasy. In the end it matters little. For the legacy of his voyage has not been the discovery of an island, it has been the creation of a space: a mysterious, unfathomable hole into which, for two millennia and more, dreams of the north have been poured. And while the desire to erase uncertainty has now wiped it from the map, Thule still exists in the cartography of the mind.

- - - - - - - -

T HE KINGS AND EMPERORS of ancient China were much concerned with the idea of immortality. For all their wealth and power they remained nonetheless human, and thus were as helpless in the face of death as even the lowliest of men. The *fangshi* were Taoist scholars whose role was something akin to that of a shaman; they were spiritual teachers, physicians, alchemists and conjurers. It was the *fangshi* to whom these anxious nobles would turn, for medicine and for advice.

According to tradition, the elixir of life, a potion bringing immortality, could be found on five islands in the East China Sea. These islands were home to the Eight Immortals, and were also the location of the 'spirit mountains', Penglai, Fangzhang, and Yingzhou. There, all plants and animals were white, and all the palaces were made of gold. The problem was that any boat approaching them would be driven off course by the wind.

The difficulties of reaching these islands did

not prevent leaders from sending out expeditions in search of them. The most famous of these journeys was led by the *fangshi* Xu Fu in 219BC and again in 210BC. Xu Fu did not travel light. On the second of these journeys he took with him 3,000 boys and girls, on sixty ships, but neither he nor they ever returned. Some believe that the travellers reached Japan, was already part of their mythology. The Fusang was a kind of mystical tree – a red mulberry and tree-of-life – that grew somewhere far to the east. Ten three-footed ravens perched in its branches, and each day one of these birds would carry the sun upon its back. Fusang was the place at which each day was born, at the foot of the Valley of the Sun.

Fusang

and that, bringing with him the technology of Chinese agriculture, Xu Fu was directly responsible for the great advancement of Japanese civilisation that took place a few decades later.

Seven centuries after this failed expedition, news of another mysterious land was heard. According to the *Book of Liang*, a Buddhist monk called Hui Shen arrived in Jingzhou in 499AD. He had come, he said, from Fusang, a place 20,000 *li* (Chinese miles) away.

Fusang was a word that people knew. It

The description Hui Shen gave of this place, however, was rather more prosaic than might be expected. The land of Fusang was indeed named after the tree, which grew there in abundance. But the tree he described was less mythical, more practical. On its branches hung pear-like fruit, and from its bark was made a kind of cloth. The inhabitants of Fusang also used the tree to make paper, for they had a written language. The monk went on to outline other aspects of this society, which had been converted to Buddhism just forty years previously. He explained the marriage and death rituals of the people, and the forms of punishment for those who had committed crimes. According to Hui Shen, Fusang was a hierarchical society, with a king – Yigi – at the top, and three levels of noblemen and aristocrats beneath. The people reared long-horned cows, as well as horses and deer, from the milk of which they made yoghurt. This was a society without money, he said, in which gold and silver had no value.

Perhaps the most remarkable thing about this account is its apparent plausibility. Nothing stands out as obviously false, and aside from its name Fusang seems unconnected to myth. There is, though, a caveat. For Hui Shen also described another land, further east, which was populated only by women. According to the monk, these women fertilised themselves by running into sacred water, and gave birth just six months later. Mothers fed their newborns from special hairs on the back of their necks – they had no breasts – and children reached adulthood by the age of four. From a biological perspective at least, this part of the tale does not stand up to scrutiny. But what about the rest?

If Fusang was indeed a real place – an island – it was certainly a long way from Jingzhou. As a unit of measurement the *li* is still in use, but its precise length has changed much over time. Fifteen hundred years ago it stood at around **400** metres, making Fusang **8,000** kilometres or **5,000** miles away. It is

hardly surprising, then, that some European scholars have located it at the other side of the Pacific. This theory came to prominence first in the eighteenth century, and was quickly accepted by some French cartographers, who placed the name where British Columbia today lies.

There are problems with this theory though, of course, not least the fact that horses were then extinct on the American continent. Other commentators have offered numerous suggestions for Fusang's whereabouts, some more reasonable than others. North-eastern Siberia, Sakhalin Island, Mongolia, Kamchatka, Hokkaido: all have been proposed as possible solutions to the riddle left by Hui Shen.

In Chinese poetry, Fusang became synonymous with the far east, in much the same way as Thule did in Europe with the north. Occasionally it was associated with Penglai, the ancient spirit mountain towards which Xu Fu set sail more than **2,000** years ago. But always it was identified most closely with Japan, or an island somewhere between the two countries.

As Japan itself absorbed elements of Chinese language and culture in the centuries after Hui Shen, it absorbed, too, this idea of itself as Fusang. One of the early names for the country was Fusō – a Japanese rendering of that very word. And just as Fusang was, in its mythical origins, the place from which each day began, so Japan imagined itself from the outside, from the west. It became, and remains, Nippon: the land of the rising sun.

St Brenda

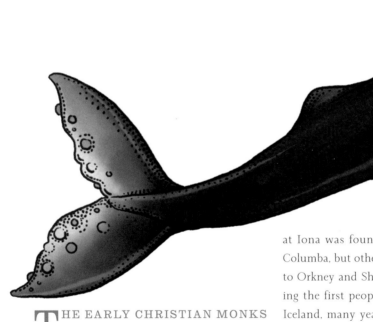

THE EARLY CHRISTIAN MONKS of Ireland sought out remote places in which to contemplate the glory of God, where prayers would be undisturbed and where faith could be strengthened through solitude and silence. They wanted isolation, and they found it on the islands of western and northern Scotland, where they began to settle in the middle of the first millennium AD. The monastery at Iona was founded in the year 563 by St Columba, but other monks went further afield, to Orkney and Shetland and beyond, becoming the first people to settle in Faroe and in Iceland, many years before the arrival of the Vikings.

Not all of these wanderers went north, though. Some, in fact, took no particular direction at all, but instead launched themselves into the ocean and let God (or the wind and currents) do the navigating. The lucky ones hit land eventually. Many others must have perished.

The best known of the travelling monks

----

n's Island

was St Brendan, who lived from 484 to 577, and was responsible for founding, among other institutions, the monastery at Clonfert in the west of Ireland. But it was not his work within the church for which he is principally remembered, it was his adventures overseas.

There are several versions of the Brendan story, each of them differing slightly from the others. Those that have survived were largely written between the tenth and twelfth centuries, but were based on earlier texts. This was a tale that was widely known across northern Europe in the High Middle Ages.

Depending on which version you read, the saint set off from Ireland in the year 512 with sixty followers, or perhaps sixteen, or fourteen. He was prompted to go by news of a glorious island – the Land of Promise of the Saints – described to him by a returning priest (or else an angel). On the journey that ensued, the monks met other holy men, as well as demons, and even the tormented soul of Judas Iscariot. They were chased by a sea serpent and a griffin; they encountered a dragon and landed on the back of a whale, mistaking it for an island. Elsewhere, they alighted on several new lands, including one known as the Paradise of Birds, and another that was home to sheep larger than oxen. The monks saw islands of smoke and fire, which surely must have been volcanoes, as well as a huge column 'the colour of silver' and 'hard as marble', consisting 'of the clearest crystal'. It could only have been an iceberg.

After seven long years, the travellers finally reached the place they had been seeking. It was, like the rest of their journey, extraordinary.

When they had disembarked, they saw a land, extensive and thickly set with trees, laden with fruits, as in the autumn season. All the time they were traversing that land, during their stay in it, no night was there, but a light always shone, like the light of the sun in the meridian, and for the forty days they viewed the land in various directions, they could not find the limits thereof.

This, clearly, was an Isle of the Blessed wrapped up in Christian language. It was a paradise on earth, to which good people would ultimately find their way. According to a young man 'of resplendent features' whom the monks met on the island: "After many years this land will be made manifest to those who come after you, when days of tribulation may come upon the people of Christ."

The story of St Brendan is a muddle of fact and fiction. Though it is impossible now to fully untangle its various threads, the tale has its roots in Irish sea sagas, or *Imrama* – particularly *The Voyage of Bran* – as well as in the Arabic story of Sinbad the Sailor and, further back, in Homer. It combines both supernatural and natural elements in a way that invites credulity yet stretches plausibility. The volcanoes, the iceberg, the giant whale, the islands of birds and of sheep: these things could easily have been seen by monks roaming the North Atlantic. Even on the final, holy island, the perpetual light could, conceivably, be that of an Icelandic summer.

But despite the northern imagery of the story, the conflation with the Isles of the Blessed and its trees laden with fruit suggested a more temperate location, and as the age of cartography arrived, mapmakers concluded that St Brendan's Island must lie further south. It appeared first in **1275** on what is known as the Hereford Map, where five islands bearing Brendan's name correspond to the Canaries, off the coast of North Africa. In the following century, many maps considered Madeira to be the holy island. Later, the Azores sometimes took the name. Cartographers, it seems, were unwilling to let go of the story (or unwilling to contradict a saint) and so they attached it to whichever island was most mysterious to them.

But at some point the mystery became more important than the geography, and the island was untangled from the known world. St Brendan's Isle began to appear on maps where no land had yet been found. And so, as the fifteenth and sixteenth centuries arrived, Brendan's name drifted westward towards the newly discovered continent. In the late sixteenth century it was there, mid-Atlantic, detached entirely from any genuine piece of land, perhaps on the assumption that it would later be rediscovered.

St Brendan may well have wandered the ocean, like other Irish monks, though nothing now could allow us to be sure. The fact he is buried at Clonfert suggests he did not go far enough to be entirely lost, but that has not prevented some from suggesting the saint may have reached the coast of North America. That theory, outlandish though it may seem, is promoted by those who argue that the Irish, not the Scandinavians, were the first Europeans to visit the New World. Fifteen hundred years later, some still put their faith in Brendan.

45

IN 711AD, A SMALL army led by Tariq Ibn Ziyad crossed from North Africa to Mons Calpe, the rock now known as Gibraltar. This was, at first, just one of a series of raids by Berbers on the Iberian peninsula, but this time things went further. Tariq's army pushed forward into what is now Spain, and

For most of the inhabitants, life did not change to any great extent. Their faith, language and laws were largely respected, as was their property. But this was not universally true, and many Christians were forced to flee the caliphate and find shelter elsewhere. Most, naturally, went north – across the Pyrenees –

The Island of

Seven

in July of that year they fought a major battle against the Visigoth troops of King Roderic. Though outnumbered, the Berbers defeated their Christian opponents, and Roderic was killed. In the years that followed, the Muslim forces were augmented by recruits from the south, and they continued their expansion through the peninsula. Virtually all of Iberia fell to the Umayyad caliphate, and much of it remained under Muslim rule for the next five centuries.

but others took more desperate measures and set sail into the ocean.

Among the escapees, it is said, were seven Iberian bishops, including the Bishop of Porto. Sailing due west, together with some of their followers, the holy men stumbled across a hitherto unknown island, far out in the Atlantic. There they burned the ships that had brought them to safety, and each bishop founded his own city: Ansalli, Ansolli,

Ansodi, Ansesseli, Anhuib, Aira and Con.

This island began to appear on maps in the early fifteenth century, and was most commonly called Antillia, or some variant of that name. On the charts of Pizzigano in 1424, and Beccario in 1435, it is shown as the largest of a group of four islands, and

is more interesting, and has several possible interpretations. Most commonly it has been assumed to mean 'the opposite island', derived from the Portuguese. This makes sense, since it sits at the opposite side of the ocean, roughly at the latitude of Gibraltar. Other theories, however, link the name to the

Cities

takes the shape in which it would remain for much of its cartographic life: a rectangle, more or less vertically aligned, with six small bays cut into its long sides and one great bite out of the southern shore. On later maps, each of these bays was given the name of one of the supposed cities.

The origins of these names are long lost, and they may simply have been the invention of a mapmaker. But the word Antillia itself

Arabic word for dragon, or to Plato's story of Atlantis. Yet another intriguing possibility is that Antillia may be a corruption of *ante Tile* – 'the island before Thule' – though that would hardly help to pin it down geographically.

Whatever its origins, Antillia was widely known in the fifteenth century. There was at least one report of a ship reaching the island, around 1430, and finding it still populated by Portuguese speakers. According to later writers, the ship returned to Europe with Antillian sand on board (sand was used in cooking boxes while at sea), which was found to contain gold. But when Prince Henry the Navigator demanded the crew return to the west to find more of that sparkling sand, they refused. Their time on the island had been unpleasant, and the strange, isolated inhabitants had left them feeling afraid. Rather than risk their lives again, or incur the prince's anger, the sailors fled.

Decades later, when Christopher Columbus was planning his ambitious voyage to China, he entered into correspondence with the physician and astronomer, Paolo Toscanelli. Labouring under the belief that the earth was rather smaller than turned out to be the case, Columbus thought he could reach Asia by sailing due west. He was seeking reassurance that such a journey was possible, and Toscanelli gave it to him. Asia could indeed be found at the far side of the Atlantic, he claimed, and if additional supplies were needed, it would be possible to stop en route at Antillia. Encouraged by this news, the explorer felt ready for his journey.

The story of the Island of Seven Cities almost certainly evolved independently from that of Antillia. One has its roots in eighth-century Iberia; the other, perhaps, hints at some early awareness of the North American continent. Though the two became entangled and then synonymous with the advent of medieval cartography, they did not remain so for long. After Columbus had sailed to the New World (an achievement he never accepted, arguing until the end of his days that he had reached Asia, as planned) the map of the Atlantic began to change. With a few notable exceptions, Antillia and the seven cities parted company. The former morphed into the Antilles – the Caribbean islands where Europeans first made landfall – but the latter drifted further afield.

On at least one map of the early sixteenth century, the seven cities were spread out over the east coast of North America, but more famously they became associated with a mythical land of plenty located somewhere on that continent. Stories of Cíbola and the Seven Cities of Gold brought the Spanish conquistadores into what is now New Mexico and Arizona in 1540. Finding only the adobe towns of the pueblo tribes, the Spaniards pushed onward to Quivira, most likely in central Kansas. Led by Francisco Vásquez de Coronado, the expedition found no gold, but it marked the beginning of European exploration in the interior of the continent, and it remains one of the founding stories of the United States.

ge of
ation

x

The Age of Exploration

The Age of Exploration

---------------   ---

THE LATE MIDDLE AGES saw a flourishing in the science and art of cartography. European mapmakers finally caught up with their Islamic counterparts, having rediscovered the geographical knowledge of Ptolemy twelve centuries after his death. From then on, advances came quickly. Exploration and colonial expansion brought new places to the map, and at home there were great minds at work on how best to portray those places to an eager audience. This was the era of Gerardus Mercator, whose world map of 1569 changed the way people saw the world. It was the era, too, of Abraham Ortelius, producer of the first modern atlas in 1570.

To look at these maps today is to see a familiar image. The planet is presented much as we know it now, with the continents more or less in their rightful places. But look closer and something else emerges. For despite the enormous leaps made by explorers and cartographers, still people were struggling to let go of what they expected to be true. Mythical islands remained stubbornly in place, and were joined by a growing legion of errors.

Maps in this period began to be printed in ever greater numbers, each one different from all those that had come before. Cartographers did not always share their sources, so when a new island appeared in the ocean it was not easy to know how it got there. From a distance of four or five centuries, it is almost impossible.

Other factors only increased the problem. A mistake on one map would be copied on others, propagating it and prolonging its life. Often, and for no explicable reason, different names would be attached to the same place. Islands migrated, drifting one way and then the other across the Atlantic. And while a single source could announce the discovery of new lands, it would usually take several non-sightings to confirm the falsity of that source. Added to all this, navigators did not find a reliable way of measuring longitude until the mid-eighteenth century, by which time their blunders were countless. The maps of this era, then, were well adorned with phantoms.

55

THE STORY OF Hy Brasil demonstrates a problem common to many of the places in this book: namely, it is hard to establish facts about phantoms. Much has been written about the island over the centuries, and much of what has been written is certainly wrong. The traditional story, repeated in countless books and articles, begins with cartography and then moves backward into folklore. It goes something like this.

From the early fourteenth century, maps produced in Genoa and then elsewhere in Europe showed an island west of Ireland, circular in shape, labelled 'Insula de Brazil', or some variant of that name. Many of these maps also showed one or two other islands elsewhere in the Atlantic with the same name, but this was merely an etymological coincidence. While these other islands – and later the South American country – were named after a kind of wood used to create red dye, the more northerly Brazil had an entirely different origin. It was derived either from the Old Irish word *bres*, meaning 'beauty' or 'strength', or else from some historical figure by the name of Breasal.

As these derivations suggest, the island was rooted in Celtic mythology. It was one of those mystical lands, like Tir na nÓg or St Brendan's Isle, that echoed back to earlier beliefs in a paradise on earth. Brazil – or Hy Brasil, as it was later known – was a place

Brasil

rarely seen. It was hidden by thick fog, and only appeared to a chosen few, once every seven years. Or so the story went.

But the line between myth and map took a long time to become settled, and it seems the belief in an island somewhere to the west of Ireland lingered for centuries. Not only did Brazil continue to appear on charts of the Atlantic, but numerous ships were sent out from Bristol in search of it. Even John Cabot, who is generally considered the first European since the Vikings to reach North America, in 1497, apparently went looking for Hy Brasil. But without success.

There were exceptions though. A few lone sailors did claim to have seen the mysterious island, including several in the seventeenth century. The best-known of these, a Captain John Nisbet of Killybegs, Ireland, gave an extraordinarily detailed description of the place, having arrived there by accident in 1674. These details were recounted the following year in a letter written by William Hamilton of Derry to his cousin in London.

According to this letter, Captain Nisbet and his crew were returning from France by sea when a thick fog descended, immersing the ship. When the fog lifted, the sailors found themselves beside an island, where cattle, sheep and horses grazed, as well as 'multitudes of black rabbits'. The following day, an old man with ten servants approached the ship and conversed with the sailors. He told them that,

until a few days previously, the island had been under the spell of a necromancer, making it invisible. Now, however, that spell was broken.

Whatever the truth or otherwise of such accounts, the island gradually began to disappear from maps, and despite a few unconvincing sightings right up into the nineteenth century, it made its last cartographic appearance in 1873 as 'Brasil Rock'. And then it was gone. An old Irish myth had been taken up by mapmakers, resulting in centuries of confusion. But finally, the confusion was cleared up.

That, at least, is the traditional story, often repeated and widely accepted. But it turns out there are some very serious problems with it.

First, there is good evidence to suggest that sailors in the late fifteenth century were using 'Brasil' as a codeword. What they were actually referring to was precisely that place John Cabot is credited with discovering: North America. Their secrecy was a means of concealing knowledge from other European powers, not just about the land itself, but about the extraordinarily rich fishery off the coast of Newfoundland.

Second, that famous account by Captain John Nisbet is a work of fiction. Not in the sense that Nisbet made his tale up – which would hardly be surprising – but, rather, that the captain himself did not exist. He was invented by the Anglo-Irish writer Richard Head, who used the island of Brasil as a setting for several of his works. The pamphlet containing this letter was a piece of satire, and quite how it came to be accepted as genuine by so many is hard to surmise. It must surely count as one of the most successful literary hoaxes of all time.

Third, and most significantly, research into the earliest recorded Irish myths has failed to turn up any mention of an island called Hy Brasil. There are plenty of tales of mysterious

sailors as code, then by writers as a fictional setting, and was finally appropriated into mythology.

None of which provides an answer as to what Brazil was doing on the map in the first place, of course. But it seems logical, once the myths and misconceptions are pushed aside, to assume that its name should have the same root as the other Brazils that appeared at the same time. Red dye, or *brazil*, was a hugely valuable

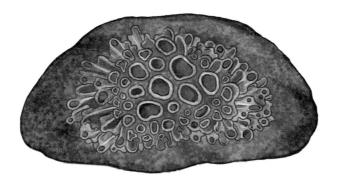

islands, both in lakes and in the sea, but neither that name nor any variation of it is used in connection with them until long after it began to appear on the map. In fact, according to a thorough study by Barbara Freitag, no reliable folkloric sources use that name until the nineteenth century. After which it became ubiquitous.

It seems that, rather than beginning in myth and ending up on the map, as long believed, Hy Brasil went the other way. It appeared on charts first, was later used by

commodity, and could be derived not just from wood but from certain types of lichen found on islands in the North Atlantic.

Perhaps the Genoese cartographers believed a source of this lichen lay to the west of Ireland. Or perhaps there were rumours even then of a new land across the ocean, where *brazil* might be found in abundance. The only thing we can say for certain is that we can never be certain. The truth of the matter is long lost. Hy Brasil is a phantom, a fiction, a myth and a mistake. It is all of these things, and in the end it is nothing.

Frisland

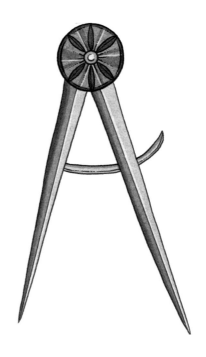

THE BRITISH EMPIRE boasted several non-existent islands at one time or another. Indeed, some of its very first acquisitions turned out not to be real. Frisland was one, and a peculiar one at that. But no less peculiar was the man who first claimed it for the crown.

Dr John Dee was a mathematician and occultist, a spy and alchemist. He was also a strangely influential figure in the court of Queen Elizabeth I. On the back of a map he presented to the monarch in 1580, Dee argued that it was not the Spanish crown that had first claim to the New World, but Elizabeth herself. North America had been discovered three centuries before Columbus, he said, by the Welsh prince Madog ab Owain Gwynedd, and was therefore British territory. But not only that. In fact, the whole of the North Atlantic region had been conquered around 530AD by none other than King Arthur. Iceland, Greenland and even the North Pole: Elizabeth was queen of them all. And what's more, Dee concluded, King Arthur 'did extend his jurisdiction and sent Colonies thither' to 'Friseland' and probably even 'the famous Iland Estotiland'.

The good doctor was not being wholly truthful. He was relying on some rather suspect sources, and adding some myths of his own. But what he didn't realise – what he had no way of knowing – was these latter two colonies did not actually exist. After all, they were not his inventions. They were there on the map. And they were there, too, on the maps of esteemed cartographers elsewhere in Europe. Neither Dee nor Elizabeth had any reason to think they might not be real.

The widespread belief in these islands, particularly in Frisland, which was to appear on maps until well into the seventeenth century, is in some ways baffling. After all, they were not discovered by any great explorer or well-respected mariner. Instead, their existence was announced by a single source, far away in Italy, who did not even claim to have been there

himself. But Nicolò Zeno, like John Dee, was a powerful man – a member of the Council of Ten of the Venetian Republic – and this wealth and political influence, along with impeccable timing, led to the rather dubious volume he published in 1558 having a major influence on the cartography of the North Atlantic.

The book, and its accompanying map, purported to be an account of a journey taken by one of Nicolò's ancestors, another Nicolò Zeno, two centuries earlier. In 1380, Nicolò the elder set off on a voyage from Venice towards England and Flanders, but he never reached his destination. Instead, on nearing the British Isles, his ship was caught in a violent storm and lost at sea for many days. When finally land was sighted, the Venetians put ashore on what turned out to be the island of Frisland. There they were met by a band of angry locals, who looked set to end the story before it had even begun. But in the nick of time, 'a great Lord' arrived on the scene. Being a nobleman, he conversed with the men in Latin, and his intervention saved their lives.

As would be expected in such a tale, the mysterious lord, Zichmni, asked Nicolò and the crew to join him in his efforts to conquer more islands in the Atlantic. And because of the skill and bravery of his new recruits, success was won without much difficulty. Nicolò, of course, was made a knight, and the Venetian invited his brother Antonio to join him in Frisland. The navigational instructions he gave

----

for the journey were presumably good, since Antonio successfully found his way north and remained in the service of Zichmni for fourteen years, before eventually returning to Venice. Nicolò himself died on the island, four years after the arrival of his brother.

While living in the north, Antonio wrote detailed letters home to a third brother, Carlo, and produced a map of the region. These letters described the travels of the two Italians, and included details about numerous other places, including Estlanda (Shetland), Engroneland (Greenland) and Islanda (Iceland), as well as less familiar locations: Groeland, Neome, Icaria, Estotiland and Drogeo. The map and letters were passed down through the Zeno family until they reached Nicolò the younger.

Had the source of cartographic confusion been these original documents, this would be a different tale indeed. The brothers Zeno would since have been cast as fraudsters or fools, and their account dismissed as nonsense. But the plot is thicker than that. In theory, there is no particular reason to doubt that the pair could have travelled in the North Atlantic. By then, the region was well known, particularly by the Norse, and islands like Faroe and Iceland were populated. What's more, these populations could certainly have included Latin speaking noblemen such as the one described in this story (often identified as Henry Sinclair, earl of Orkney). But in this case, however, an additional layer of confusion was added to the story, not by the elder Zenos but by Nicolò the younger.

No one except Nicolò seems ever to have laid eyes on the original map or letters, and the book he produced in 1558 was not even a direct transcription of the originals. In fact, Nicolò claimed that, as a child, he had read and, rather inexplicably, destroyed the letters. What he wrote down much later was partly from memory and partly from the remaining fragments he could piece together. The map, similarly, was rotten by the mid-sixteenth century, and must have required some imaginative interpretation to bring it back to life.

What these alleged facts leave behind, therefore, are several possibilities. Either the story is true in essence, with allowances made for Nicolò's poor geography and the limitations of his source material. Or else the brothers invented their own myth, concocting Zichmni and his Atlantic islands in order to put themselves, quite literally, on the map. Alternatively, and most plausibly, it was Nicolò the younger who was the mythmaker, and his book was a partial or total invention.

Had this tale been told by a less wealthy or powerful man, and had it not been taken up by the likes of Mercator and Ortelius, Frisland would have sunk back into the ocean immediately and would never have been seen again. Instead, it appeared on maps of the north for over a hundred years, and continues to be debated today. It fooled cartographers, magicians and monarchs.

ONLY ONE BRIEF BUT intriguing account of Davis Land exists, and it comes from the memoirs of a Scots-Irish surgeon and pirate by the name of Lionel Wafer. Though he began his career as a conventional ship's doctor, Wafer, like many young men in the late seventeenth century, was drawn towards greater adventures. In 1679, he was recruited in Jamaica by the buccaneer Edmund Cook, and the following year, together with more than 300 others, he crossed the Isthmus of Darien to raid Panama from the Pacific coast.

Wafer sailed with William Dampier, another pirate who would later gain some respectability as an explorer and naturalist. The pair were among dozens who returned to Panama in 1681, with the aim of reaching the Atlantic again. But things did not go according to plan. During the crossing, Wafer was seriously injured when a plate of gunpowder exploded beside him, scorching and tearing the flesh from his leg, exposing the bone. A few days later, things turned even worse when a group of slaves escaped, taking with them not only guns and money but also the medicine necessary to treat the wound. Unable to continue, the pirates left Wafer behind in the hands of the local Cuna people, who saved his life.

During his time with them, he learned their language and familiarised himself with the customs and practices of the indigenous people. In turn, he taught the Cuna some of his own medical knowledge. Eventually, the following year, Wafer made it through to the Atlantic coast, where he found his former shipmates again. Dressed and painted like a native, and sporting a large nose ring, it took them almost an hour to recognise him.

Davis

For several years, Wafer and Dampier continued to sail as buccaneers, first in the Caribbean, then around Africa and back into the Pacific. Though both men wrote memoirs that sought to minimise their involvement in illegal activities – focusing instead on the land, people and wildlife they encountered – both were up to their necks in it, taking part in countless raids and attacks on Spanish vessels. After one of these assaults, Wafer transferred

Land

to a stolen ship, renamed the *Batchelor's Delight*, first under Captain Eaton and later Edward Davis. Together they explored the islands of the eastern Pacific and the coast of South America.

In October 1687, the men were heading south from the Galápagos Islands, intending to reach Juan Fernández (the island from which, a decade later, William Dampier would rescue Alexander Selkirk, the castaway whose story inspired the tale of Robinson Crusoe). But en route the crew experienced a massive earthquake that confused and terrified the sailors. They continued as planned, but what they found was not what they expected. According to Wafer's memoir:

Having recover'd our Fright, we kept on to the Southward. We steer'd South and by East, half Easterly, until we came to the Latitude of 27 Deg. 20 Min. S when about two Hours before Day, we fell in with a small, low, sandy Island, and heard a great roaring Noise, like that of the Sea beating upon the Shore, right a Head of the Ship. Whereupon the Sailors, fearing to fall foul upon the shore before Day, desired the Captain to put the Ship about, and to stand off till Day appeared; to which the Captain gave his consent. So we plied off till Day, and then stood in again with the Land; which proved to be a small flat Island, without the guard of any Rocks. We stood in within a quarter of a Mile of the Shore, and could see it plainly; for 'twas a clear Morning, not foggy,

nor hazy. To the Westward, about 12 *Leagues by Judgement, we saw a range of high Land, which we took to be Islands, for there were several Partitions in the Prospect. This Land seem'd to reach about* 14 *or* 16 *Leagues in a Range, and there came thence great Flocks of Fowls. I, and many more of our Men would have made this Land, and have gone ashore at it; but the Captain would not permit us. The small Island bears from* Copayapo *almost due E.* 500 *Leagues; and from the* Gallapago's, *under the Line,* 600 *Leagues.*

And that was that. One small, flat island, and another, or perhaps others, to the west. Captain Davis' failure to explore his discovery makes it impossible to identify, but after Wafer's account was published in 1695, numerous sailors were encouraged to search for it. After all, the pirates knew this region better than anybody else alive, and if the land was unknown to them it was certainly unknown to others.

For more than one hundred years, people sought Davis Land without success. Eventually, some assumed the men had stumbled upon Easter Island (and they would, indeed, have been the first Europeans to do so). But Easter Island is alone, and that explanation, like all others, seems not quite good enough.

The following year, Lionel Wafer was arrested and briefly imprisoned for piracy, before sailing to England and abandoning forever his buccaneering ways. At that point he might easily have disappeared from history, were it not for the fact that his expertise proved suddenly to be in great demand.

Wafer's knowledge of the Darien Isthmus was sought out first in 1697 by John Locke and the Council of Trade and Plantation. Then, in secret, he was contacted by the directors of the Darien Company, and his advice proved to be influential in the siting of the new Scottish colony on the isthmus, based around a settlement called New Edinburgh. But Wafer's enthusiasm for the project was not well-founded.

Around twenty per cent of all the money in Scotland was invested in the Darien Company's venture, and all of that money was lost. The scheme was disastrous, and the consequences of the failure were huge. The near-bankruptcy of the nation that resulted was in large part responsible for the birth of the United Kingdom in 1707, when Scotland rather grudgingly signed the Act of Union with England. It was a union that Lionel Wafer did not live long enough to see.

IN EDGAR ALLAN POE'S novel of 1838, *The Narrative of Arthur Gordon Pym of Nantucket*, the eponymous sailor travels to the far south aboard a ship called the *Jane Guy*. Its captain is seeking fur seals and exploring the remote waters beyond the Cape of Good Hope. He visits the Prince Edward and Crozet Islands in the southern Indian Ocean, then goes onward to the Kerguelen archipelago, two thousand miles from the nearest human habitation – a place known, without affection, as the Desolation Islands.

The *Jane Guy* then turns west, sailing into the South Atlantic. It passes Tristan da Cunha and continues towards the far side of the ocean. The captain now has another destination in mind. He is looking for the Auroras, a group of three small islands midway between the Falklands and South Georgia. His aim, says Pym, is to 'settle the question so oddly in dispute'. That question: do the Auroras really exist?

We kept on our course, between the south and west, with variable weather, until the twentieth of the month, when we found ourselves on the debated ground, being in latitude 53 degrees 15' S., longitude 47 degrees 58' W.— that is to say, very nearly upon the spot indicated as the situation of the most southern of the group. Not perceiving any sip of land, we continued to the westward of the parallel of fifty-three degrees south, as far as the meridian of fifty degrees west . . . We then took diagonal courses throughout the entire extent of sea circumscribed, keeping a lookout constantly at the masthead, and repeating our examination with the greatest care for a period of three weeks, during which the weather was remarkably pleasant and fair, with no

The

haze whatsoever. Of course we were thoroughly satisfied that, whatever islands might have existed in this vicinity at any former period, no vestige of them remained at the present day.

Pym's experience drew on and reflected the accounts of other, real-life sailors who had gone looking for the Auroras. In **1820**, James Weddell thoroughly searched the area and found nothing; and two years later Benjamin Morrell claimed to have done the same thing. These reports and others led, ultimately, to the removal of these islands from the map.

But the Auroras are somewhat different seen numerous times, and by very credible witnesses, over a period of several decades.

The first known account of their location came in **1762** from the whaling ship *Aurora*, which gave them their name. The same ship saw them again twelve years later, though not before the *San Miguel* confirmed their existence in **1769**. They were seen a fourth time in **1779**, then twice more in **1790**. And

Auroras

from most ex-isles, and expunging them was not the simple matter it might otherwise have been. For they were not the product of a single, unreliable sighting, and nor was their location inadequately recorded. In fact, the islands were if there were any doubts at this stage as to the reliability of these sightings, they must surely have been assuaged by what happened next. For in **1796**, a Spanish research vessel, the *Atrevida*, was assigned the task of precisely

----

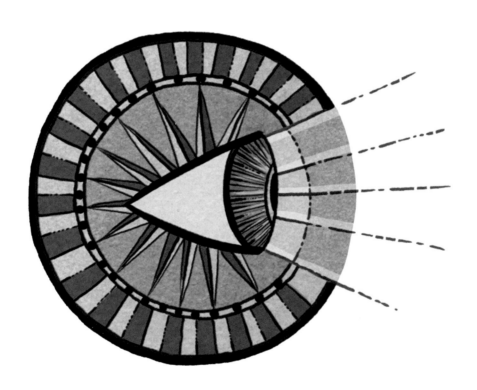

locating and surveying the islands, which she did, apparently without difficulty.

According to the log of the *Atrevida*, the largest of the three Auroras comprised 'a great mountain in the form of a pavilion (or tent) divided vertically into two parts, the eastern extremity *white*, and the western very *dark*; on which latter side was a snowy band'. The following day, the second of the islands was recorded, and was found to be 'also covered in snow, but not so high as the former one'. Three days later they found the third: 'a large rock, making in sharp pinnacles, but formed like a saddle-hill'. The exact location of each was checked using chronometers that had been tested only a few days previously.

The Auroras are surely unique among non-existent islands, having not only been sighted numerous times but also surveyed by a highly skilled and qualified crew. It is no surprise, then, that nautical charts from the late eighteenth century onwards included them, and that most mariners chose to avoid the area as they headed for Cape Horn. Such dangers to shipping had to be taken seriously.

But aside from two rather less convincing sightings, in 1820 and 1856, the Auroras were never seen again. In half a century or so their status changed from certain to doubtful to gone, then finally to almost forgotten. Few islands have had such a precipitous decline. So what on earth happened to them?

James Weddell – an authority on the region, and a captain of tremendous ability – blamed nothing more sinister than human error. After completing a thorough search of the area in 1820, he concluded 'that the discoverers must have been misled by appearances; I therefore considered any further cruise to be an improvident waste of time'. The most likely cause of confusion, he speculated, was the Shag Rocks, a group of six islets at roughly the same latitude as the Auroras, but more than 100 miles to the east. These lonely skerries, or else a trio of rock-like icebergs, must surely be to blame.

It is a logical conclusion, and one that is impossible to disprove. Yet even Weddell found himself hard-pressed to explain how experienced mariners could have made such a blunder. And not just once, but several times. The *Atrevida* carried the very best of navigational equipment, and had the most able of sailors and scientists on board. Even ignoring the fact that their description of the Auroras does not match that of the shark-tooth Shag Rocks, an error of 100 miles, or six degrees of longitude, seems highly improbable. Perhaps, Weddell suggested, the Spaniards were simply unused to the 'cold and tempestuous seas, encumbered with ice'. Perhaps he was right.

There is, in the end, no explanation that seems entirely adequate to account for the appearance, then disappearance, of the Auroras. They remain a mystery, and among the most inexplicable of phantom islands.

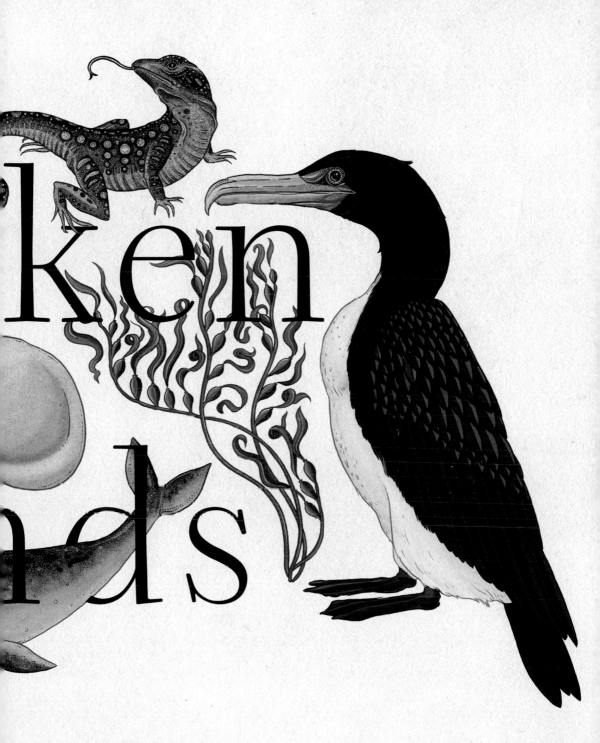

Sunken Lands

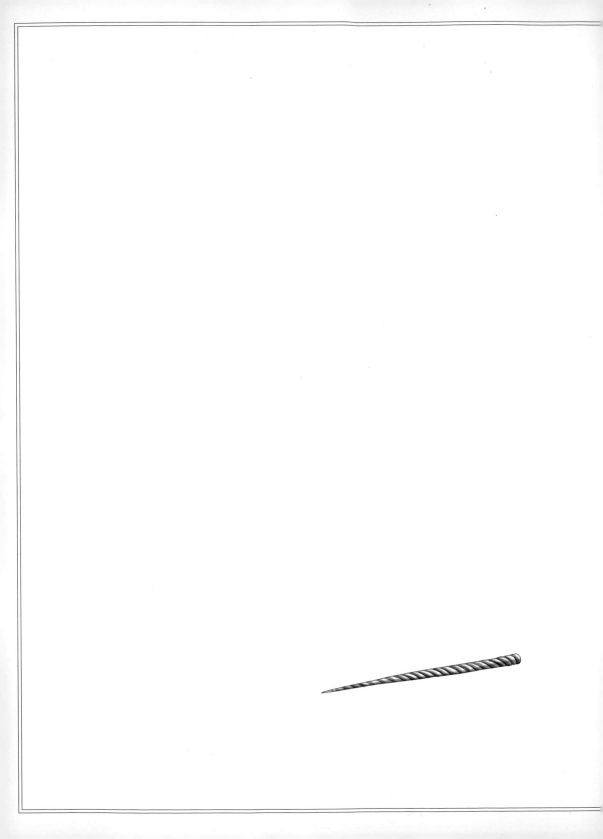

Sunken Lands

T HE STORY OF NOAH and his ark is just one of many ancient flood myths recorded across Europe, Africa, Asia and the Americas. Whether these tales represent cultural memories of rising sea levels or tsunamis is impossible now to know, but what is clear is that submergence has long been an important motif in the stories people tell. It is perhaps unsurprising then that Atlantis has become the archetypal ex-isle: a strange land lost to the ocean. No matter that it was conceived as fiction, it now serves another purpose entirely: as a kind of sponge for human fantasy.

Several of the islands in this book have, at one time or another, been thought to have sunk. In the early days of European exploration, sailors were generally assumed to be reliable witnesses, so when new lands were reported they would be faithfully added to the map. And if those new lands could not be found again, natural causes for the disappearance would be assumed before error or fraud were suspected.

One of the difficulties in determining the truth of the matter, though, is that sometimes islands *do* sink. In some parts of the world, land that has long seemed stable can disappear almost without warning. Up until the fourteenth century, Gunnbjörn's Skerries were a stopping off point for Norse travellers between Iceland and Greenland. For a time they were even farmed. Then, following a volcanic eruption in 1456, they were gone. Another Icelandic island, Geirfuglasker, sank in similar circumstances in 1830. Inaccessible to hunters, it had previously been the last safe breeding ground for great auks, which were flightless and clumsy on land. When it disappeared the great auks were forced to breed elsewhere, and within fifteen years they were extinct.

But question marks remain over some disappearances. Tuanaki, in the Cook Islands, was visited in 1842, but two years later a group of missionaries failed to find it. There were reports that some Tuanakians had managed to flee to Rarotonga in the south before their home sank, and the story of its submergence is generally accepted to be true. But it is very hard, in retrospect, to be sure. The idea of a drowned island is somehow both irresistible and unbelievable.

THOUGH UNDOUBTEDLY THE most famous of all ex-isles, Atlantis is not strictly speaking an un-discovered island; it is a fictional island, invented by Plato for allegorical purposes. The story – told in two of his dialogues, *Timaeus* and *Critias* – was never supposed to be taken literally. But while scholars today are almost universally agreed on this point, there was more than enough ambiguity in those works to fuel two thousand years and more of speculation and pseudoscience.

The island appears initially in the introductory conversation of *Timaeus*, written around 360BC. That dialogue, which goes on to consider the origins, purpose and properties of the universe, was followed by the un-

finished work *Critias*, in which the tale of Atlantis is expanded. The details contained in these two dialogues are a mixture of the fantastical and the almost believable.

Atlantis was a large island, they state, located somewhere in the Atlantic Ocean. The dynasty of kings who ruled there were descendants of Poseidon, the god of the sea, but their divinity had become diluted 'by frequent admixture with mortal stock'. They had become proud, over-ambitious and degenerate. Nine thousand years before Plato's day, these kings were waging war against the people of the Mediterranean, and had already conquered much of the region.

Critias provides a great deal of information, not just about the geography and politics of the island, but about the architecture of its capital city, the design of its irrigation ditches and the characteristics of its ritual sacrifices. *Timaeus*, on the other hand, concerns itself with the final, key elements of the story: the bloodthirsty Atlanteans were defeated by the warriors of Athens. Then the island itself was destroyed in a single day and night, by earthquake and by flood, along with Athens itself.

The story of Atlantis is not presented within the dialogues as being fiction. Quite the opposite: it is told as though it were historical fact. But that does not mean that

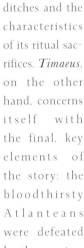

it *is* historical fact. Even leaving aside the mythological elements of the tale – and, also, the sheer impossibility of what is described – it is difficult to avoid the conclusion that what we are dealing with is elaborate allegory, not ancient history.

But that has not stopped people claiming otherwise. A few early commentators – most notably Strabo, two thousand years ago – considered the possibility that the story might be true. But they were the exception. It was not until Europeans began to explore the American

an integral part of our cultural landscape.

Some of the speculation surrounding the Atlantis myth has been vaguely scientific. Researchers have attempted to tie the story to real geological events. Some have argued that the Minoan eruption, an enormous volcano that devastated the island of Thera – today's Santorini – around 1600BC, provided the historical inspiration. Others have pointed to Doggerland, which once joined the east of the British Isles to mainland Europe, and which was finally submerged more than eight thousand

Atlantis

continents in the sixteenth century that the idea of Atlantis as geographical fact, or even as key to some deeper secret about the world, really began to take hold. Since then, hundreds of books have been published on the subject; television programmes and films have been made. This sunken island, which played only a minor role in Plato's work, has risen to become

years ago; or to Helike, a Greek city destroyed by a tsunami during Plato's own lifetime.

But alongside these theories has been another, less credible but no less persistent, thread of investigation. Though the claims are diverse, the central idea is generally this: Atlantis, wherever it was located, was the birthplace of civilisation, and a lost utopia. In the words of

Ignatius L. Donnelly, whose **1882** book *The Antediluvian World* is certainly one of the most influential of its kind, Atlantis was once home to 'a great, wise, civilized race'. A few of these people migrated to Egypt, to Europe, to North and South America, and carried with them their technology – including writing – as well as stories of their kings and queens, who came in time to be thought of as gods. Later, a great flood drowned the island and its people. That flood was the same one described in the Bible and in other, similar stories on both sides of the Atlantic.

Though Donnelly's claims are certainly crazy, he is far from a lone voice. Today, if you want to explore the astonishing limits of human credulity, searching online for websites about Atlantis is a good place to start. There one can learn that proof of the story has been found within the Great Pyramids of Egypt (which are, some claim, not tombs but ancient power stations or radio transmitters). Others argue that Atlantis was situated in the Caribbean, and was destroyed by a meteor shower; or else it was in Morocco, on the edge of the Sahara Desert. One persistent theory is that Antarctica was Atlantis, but a shift in the earth's crust caused it to move southwards. Some have even suggested

it was a planet that once orbited our own.

In fact, you can discover almost anything you want to discover about Atlantis, and pretty much every word of it is nonsense. Theories rise, then topple beneath the weight of their own silliness, to be replaced by several other, equally silly ideas. To investigate them is to be faced with an impenetrable forest of absurdity, into which no one ought to wander.

The paradox of course is that, at their most basic level, these theories are not so far away from what Plato originally intended, for he too was imagining a lost, perfect society. The difference is that his utopia was ancient Athens,

'conspicuously the best governed [state] in every way, its achievements and constitution being the finest of any in the world'. Atlantis, for Plato, was merely an imaginary foe.

The classical scholar Desmond Lee called the Atlantis story 'the first exercise in the art of science fiction'. In it, the familiar and the fantastical are tied together to illuminate elements of the real world. The island was a pawn in a rhetorical game; it was created in order to be sunk. That it is still with us, more than twenty-three centuries later, is surely the most extraordinary thing about Atlantis.

----

small fishing vessel, of which the *Emmanuel* was one). There were, at least to begin with, no doubts raised about its existence, but it was certainly elusive. Buss was seen again in 1605 by Captain James Hall, but four years later Henry Hudson could not find it. Nor could anyone else, in fact, until Zachariah Gillam in 1668, then Thomas Shepard in 1671.

Despite its three previous sightings, Shepard was the first sailor who claimed to have actually set foot on Buss. Not only that, he explored and mapped the island, showing it as roughly diamond-shaped, with two narrow inlets on the southwestern coast and a broad range of mountains in the northwest.

On his map, and in the accompanying description, Shepard named several of the island's natural features after directors of the newly incorporated Hudson's Bay Company, including Rupert's Harbour (after Prince Rupert), Viner's Point, Griffith's Mount and Robinson Bay. This was a canny act of flattery, which was dangled alongside descriptions of the island's bountiful resources: fish and whales in abundance. This was bait that could not be resisted.

The Hudson's Bay Company sought a charter from King Charles II in 1673, which was granted two

years later, giving them a monopoly on trade around the island of Buss. It cost £65, but did not prove to be their best investment. When Shepard was sent out to find the island again, and to secure the fortune that he and the company had been expecting, he couldn't do so. The island was not there.

Shepard's map and his description of Buss had been an invention. Assuming, perhaps, that the previous sightings could not all be wrong, he had taken a gamble and lied about his own part in the story. It was a bold move, which could easily have paid off. Shepard would surely have been as surprised as anyone to learn that Buss did not exist.

By the middle of the eighteenth century, the general assumption was that the island had sunk; some geological catastrophe had occurred, and Buss had been consumed by the ocean. There was logic to the theory, given the plausibility of the previous reports. Maritime charts sometimes marked the area as a reef or shoal – a potential hazard to shipping – and several captains reported shallow water in the area where Buss might once have been. But in **1818**, Sir John Ross of the *Isabella* (who later that year would discover the non-existent Croker Mountains, in the Canadian Arctic) could find no water shallow enough to be a sunken island. Thereafter, the theory dissolved.

The most likely explanation for the appearance and disappearance of Buss is simply that its original discoverers, on the *Emmanuel*, were lost. They had been blown off course by the storm, and what they saw was Cape Farewell, or some other part of southern Greenland. Certainly, there is little in the account of Thomas Wiars that would make this implausible. Later, Captain Hall might have been deceived by ice or by an optical illusion, and he saw what the charts told him should be there· an island.

Perhaps Zachariah Gillam in **1668** was likewise confused, though no first-hand account of his sighting was ever published. It is worth noting, however, that on that particular Atlantic crossing – an historic expedition that had led, ultimately, to the founding of the Hudson's Bay Company – the chief mate aboard Gillam's ship was none other than Thomas Shepard.

Sarah Ann Island

WARS HAVE LONG BEEN fought over precious resources. Oil has been at the heart of numerous recent conflicts, and water is almost certain to fuel future ones. In the twentieth century, diamonds helped to stoke civil wars in Africa, and disputes over fishing grounds in the North Atlantic led to the three – fortunately bloodless – 'cod wars' between Britain and Iceland, from the 1950s to the 1970s. But a hundred years earlier it was a different resource altogether that brought conflict to the western coast of South America: bird droppings.

High in nitrogen and phosphates, guano, as it is known, can be used to produce agricultural fertilisers, and in the mid-nineteenth century a lucrative trade was developing, with Peru as the main exporter. On arid islands, where birds had previously been undisturbed by people, the guano could be found in extraordinary depths, up to 150 feet. Before long, hundreds of thousands of tons were being removed each year.

From early on, the United States was keen to cut in on the guano trade, but its first attempt was clumsy and entirely unsuccessful. The country got itself involved in an angry confrontation in 1852, when it brazen-

Lemuria

or

Kumari Kandam

IN THE MID-NINETEENTH centu-
ry, the British zoologist Philip Sclater was
faced with a puzzle. A well-respected scientist, who would go on to become secretary of
the Zoological Society of London, as well as
founding editor of *Ibis*, the journal of the British Ornithologists' Union, Sclater specialised
in zoogeography. He studied the distribution
of animals and birds around the world, and
was the first to divide the
planet into the biological regions which today are called
ecozones.

Though he wrote most
extensively on the birds of
South and Central America,
Sclater's interests were
wide-ranging and included
the wildlife of Madagascar,
which he described as 'one of
the most anomalous faunas
existing on the world's
surface'. The anomaly that
particularly interested him
was the lemur, a primate endemic to the
island but with relatives both on the African
continent and in India.

The connection with Africa was easily
explained – the island is only **250** miles
offshore – but India was a more troubling
question, for a whole ocean divides it from
Madagascar. This was the puzzle facing Philip
Sclater, when, in **1864**, he wrote a short essay

for *The Quarterly Journal of Science*, in which
he suggested one possible explanation:

> *that anterior to the existence of Africa*
> *in its present shape, a large continent*
> *occupied parts of the Atlantic and*
> *Indian Oceans stretching out towards*
> *(what is now) America to the west, and*
> *to India and its islands on the east;*
> *that this continent was*
> *broken up into islands,*
> *of which some have*
> *became amalgamated*
> *with the present continent*
> *of Africa, and some,*
> *possibly, with what is*
> *now Asia; and that in*
> *Madagascar and the*
> *Mascarene Islands we*
> *have existing relics of this*
> *great continent, for which*
> *. . . I should propose the*
> *name Lemuria!*

In the nineteenth century, the notion of a
sunken continent was not a particularly contro
versial one. It was a logical answer to a difficult
question, and Sclater was not the first to suggest
it. Nor would he be the last. Though some of his
contemporaries, including Charles Darwin, were
unconvinced, others took the idea and went
even further with it.

It was a German biologist, Ernst Haeckel,

who first suggested Lemuria might be the birthplace of mankind. Like Sclater, he was well respected and influential, but prone to occasional leaps of logic and rather dodgy theories about race. In his book, *The History of Creation*, published in English in 1876, Haeckel proposed that the evolutionary step from ape to human could have occurred in the now sunken land, thus explaining the shortage of 'missing link' fossils elsewhere. On a map contained in that book, the various peoples of the world were shown emanating from Lemuria, which was given the additional label, 'Paradise'.

Both Sclater and Haeckel were scientists. They recognised their hypothesis lacked solid evidence, and neither tried to claim Lemuria as fact. But others did. The idea, once planted, took on a life of its own.

Haeckel's writing was influential not just among European racists, but also among the very same people who were then producing and con-suming books about Atlantis. Indeed, many of these books began to link the two locations, with the Lemurian people sometimes identified as ancestors of the Atlanteans.

Some of the claims made about Lemuria were, if anything, even more preposterous than those linked with Plato's story. Helena Blavatsky, co-founder of the occultist Theosophical Society, argued in her book *The Secret Doctrine* that the Lemurians had been hermaphrodites with four arms, who laid eggs rather than giving birth. They were, she wrote, one of the seven 'root races'

of the world. A later writer offered a somewhat different description, suggesting that the people of the lost land were twelve to fifteen feet tall with thick scaly skin, and that they kept large reptiles as pets. Another theory – still popular today – is that descendants of the Lemurians live inside Mount Shasta in northern California.

But it wasn't just Western fantasists who took an interest in Haeckel's hypothesis. In India, the traditional stories and literature of the Tamil people included tales of a land that disappeared beneath the ocean, possibly destroyed by tsunamis. These stories had not previously been given much cultural status, and there was no clear indication within them of how large the land had been. But when the

idea of Lemuria found its way into Indian textbooks at the end of the nineteenth century, the traditional tales were given unexpected credence.

The Tamil people have an ancient language

and culture, centred around Tamil Nadu in south-east India and on the island of Sri Lanka. The early twentieth century saw a revival of this culture, and a determined effort to push back against the dominance of other Indian languages. The notion of a lost homeland (either an island or a larger extension of the Indian subcontinent) was propagated by revivalist writers, and gradually

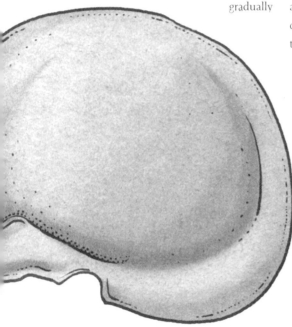

Lemuria – renamed Kumari Kandam – became a central part of contemporary Tamil mythology.

A peculiar mirror image then developed between Western and Eastern interpretations of this story. While some in Europe and North America used the idea of a Lemurian 'root race' to justify their belief in Aryan superiority, Kumari Kandam was specifically used in India to promote another falsehood: that the Tamil language and culture was the oldest and purest in the world. Some even claimed that all other peoples were descended from Kumarian Tamils.

By the mid-twentieth century, science had moved on. The theories of continental drift and, later, of plate tectonics provided a more convincing answer to the questions that had troubled biologists and geologists a hundred years before. India and Madagascar were indeed connected, it turned out, but not by a land bridge, and certainly not as recently as Ernst Haeckel imagined. The split took place more than 80 million years ago, a very long time before anything like a human being had evolved. The idea of Lemuria, and of other sunken continents such as Mu in the Pacific, was swiftly abandoned by all but the most gullible.

In India, though, it proved much harder to let go of the belief. Kumari Kandam had become important to people's sense of cultural identity, and so it remained. Right up until the early 1980s the story of a lost homeland was taught in some Tamil schools as though it were fact, and still today many believe it has science on its side. Philip Sclater's hypothesis may have been disproved, but the land he imagined has not yet disappeared entirely.

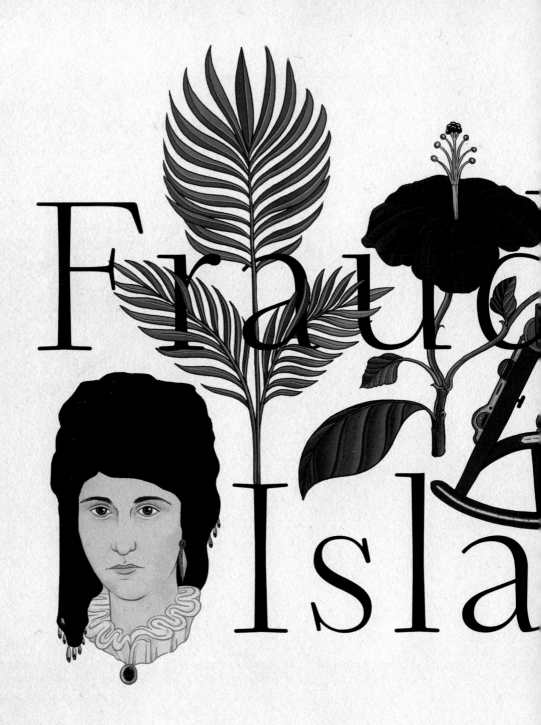

x

Fraudulent Islands

Fraudulent Islands

AMONG THE MULTITUDE OF non-existent islands that have appeared on maps over the past few centuries, the vast majority are the result of mistakes. They are accidental phantoms, caused by imperfect navigation, optical illusions or poor recording by mariners and cartographers. Sometimes, though, there is no accident at all. Islands are invented deliberately, often creating inordinate confusion as a result.

There were many reasons for mariners to lie about what they had discovered, especially if the chances of being found out were slim. Fame and prestige could be found by the most creative liars; the public were often eager to learn of new discoveries overseas, and a captain with a good imagination could do well. But there was money, too, to be made in this business. Investors would only support explorers if they thought it worth their while, so there was plenty to be gained by lying. Rich businessmen and politicians were also keen to be immortalised, and what better tribute could there be than an island bearing their name?

Just as it was hard to identify errors on the map, lies could take a long time to uncover, and they could cause a lot of problems in the process. In **1910**, the International Date Line had to be redrawn after it was discovered that Byres Island and Morrell Island – around which it had previously diverted, northwest of Hawaii – did not actually exist. They were the inventions of Benjamin Morrell, a nineteenth-century sealing captain, whose reputation for fantasy was widely known in his own lifetime. Morrell's semi-fictional memoir, *A Narrative of Four Voyages*, included accounts of his searches for other doubtful islands, including the Auroras. And though he apparently succeeded in finding Bouvet Island – a remote speck in the South Atlantic, which many had previously thought to be a phantom – the American couldn't resist inventing a couple of his own. He was far from the only one.

Isle P

THE LOCATION OF international borders has often been the cause of conflict, and sometimes of confusion. Perhaps only once, however, has a border been marked by a fictitious island. And extraordinary as it may now seem, that border was between the British colonies in Canada and the newly founded United States of America.

When the American Revolutionary War ended in 1783, confirming for the first time the existence of the United States as a sovereign, independent nation, the limits of that new nation had to be defined. The necessary negotiations for this took place in Paris, and were led, on the American side, by Benjamin Franklin, John Jay and John Adams, and on the British by David Hartley. The result of these talks was the Treaty of Paris, signed on

helipeaux

3rd September of that year, which set out the terms of the peace brokered between the new country and the old. While the first article of this short treaty was an official relinquishment of all claims by King George III on the territory of the United States, the second – a single paragraph of just over five hundred words – described the boundaries of that territory.

By the terms of the treaty, the western border of the country was marked by the Mississippi River, and the southern by the 31st parallel, and by a series of smaller rivers, separating it from Spanish-held Florida. In the north, things were more complicated, both geographically and politically, and most of the second article was concerned with defining the border as it meandered from the coast of Maine westward through the Great Lakes.

According to the text, the line ran through the middle of Lake Ontario, Lake Erie and Lake Huron. Beyond that, it went 'through Lake Superior northward of the Isles Royal[e] and Phelipeaux, to the Long lake; thence through the middle of the said Long lake and the water communication between it and the Lake of the Woods'. So far, so clear.

Though the declared intention of the Treaty of Paris was to secure 'perpetual peace and harmony' and to ensure that 'all disputes which might arise in future on the subject of the boundaries of the said United States may be prevented', it achieved no such thing. Disputes were inevitable, and they continue

right up to the present day (Machias Seal Island and North Rock in the Gulf of Maine are even now claimed by both the USA and Canada).

Perhaps the first major test for the border, however, occurred thirty years after it was drawn, when Britain and the US once again went to war. That three-year conflict, beginning in 1812, did not result in any boundary changes, but it did highlight the need for more clarity, particularly in the north. Consequently, after the war two commissioners were given the task of surveying and marking the border between Lake Huron and the Lake of the Woods – one of the regions in which clarification was deemed to be necessary. This turned out to be a more difficult task than expected.

When the surveys were carried out in the early 1820s, problems were found with the wording of the original treaty. For one thing, it was not at all clear what was meant by the 'Long lake'. Was it one of the many long lakes in the area? And if so, which one? Or else did it refer to an elongated inlet of Lake Superior, or even a river that had been incorrectly denoted? Nobody was sure.

Another problem, equally baffling, was that the surveyors were unable to find one of the islands mentioned in the treaty. Though the position of the border within the lake was unaffected, the failure to locate Isle Phelipeaux must nevertheless have been a surprise and a cause for concern. At the very least, it showed that a new chart was needed for the north of the country.

The map that had been used during the negotiations in Paris, forty years previously, was the best and most detailed one available at the time. First published in 1755, it was known as the Mitchell Map; and though a superb piece of cartography, and certainly one of the most important maps in American history, it did contain a number of errors. John Mitchell, who produced the map, did not invent Isle Phelipeaux, but it was undoubtedly an invention. His source for the Great Lakes region was another map, of eleven years earlier, by the French geographer Jacques-Nicolas Bellin. But Bellin too was innocent.

Finding the culprit in this cartographical fraud might have proved impossible, had it not been for the fact that the guilty party had gone too far. On further exploration of the area it was discovered that there was not one but four fictional islands in Lake Superior, and in the end it was their names that gave the game away. Alongside Isle Phelipeaux were the equally non-existent isles of Pontchartrain, Maurepas and St Anne – names which all pointed in a very particular direction.

Phelipeaux (or, more correctly, Phélypeaux) was the surname of Jean-Frédéric, secretary of state for maritime affairs in France from the 1720s up until the end of the 1740s. As well as a politician, Phélypeaux was also a

count, whose estates were Pontchartrain and Maurepas. The family's patron saint, rather predictably, was Anne.

Naming islands, rivers, mountains and other geographical features after rich men back home was a popular activity among explorers. After all, there are few things such men appreciate more than flattery, and when someone is looking for money to fund future travels such considerations become important.

As it happens, Jean-Frédéric Phélypeaux did indeed patronise foreign exploration, and among the recipients of his generosity was Pierre François Xavier de Charlevoix, a Jesuit priest, historian and traveller, who had spent time in the Great Lakes region in the 1720s, and who worked for the French crown trying to find a route to the Pacific. The clinching detail that ties Charlevoix to this particular fraud is his book, *A History of New France*. It was for that book, published in 1744, that Bellin's map, on which the islands first appeared, was commissioned.

That they later turned out to be fictional mattered little, for by the time the fraud was uncovered all those involved were dead. The border complications resulting in part from the lies of Charlevoix and the errors of Bellin and Mitchell were resolved in the Webster–Ashburton Treaty, which was signed in August 1842. The islands of Phelipeaux, Pontchartrain, Maurepas and St Anne then disappeared from the map forever.

Java

ISLANDS ARE USUALLY discovered by those who leave their homes and sail in search of land and adventure. They are found by those who go looking. But on one occasion, it was the island that came looking for people, arriving in the shape of a mysterious woman in a black turban.

It was the evening of Thursday, 3rd April 1817, in the village of Almondsbury, near Bristol. On the doorstep of a cottage stood a woman of about twenty-five, who had just knocked on the door. In addition to her turban, she wore a black dress with a frill at the neck, and a red and black shawl around her shoulders. She had a pair of leather shoes and woollen socks, and had clean, soft hands, like those of someone unused to work. It seemed she was looking for a place to spend the night, but the words she spoke were not English. They were, in fact, entirely unrecognisable.

On the supposition that the woman might be a vagrant, or even a spy, she was taken first to the overseer of the poor and then to the house of Samuel Worral at Knole Park. Mr Worral was the local magistrate, and he also employed a servant who could speak several European languages. But the servant, when called to assist, was equally puzzled. He could make nothing of what the woman said. Without any papers or belongings, and without a language that could be understood, she was a blank person, a nobody. And had it not been with recognition, as was furniture imported from that country (though she did not appear to be Chinese). Many of the woman's habits, too, were rather odd. She seemed reluctant to sleep in a bed, preferring the floor; she drank only tea and water, and would not eat meat; she recited prayers with one hand covering her eyes. Beyond that, though, little could be learned.

To begin with, there were those who

a s u

for the intervention of Mr Worral's wife, her fate would most likely have been prison or deportation. But for now, she was lucky.

In this woman, so far devoid of an identity, Elizabeth Worral saw a mystery she wanted to unravel, and over the coming days she began to piece together such clues as could be uncovered. These did not amount to much. The woman's name, it seemed, was Caraboo. At least that was the word she used while pointing to herself. Images of China were met

doubted that Caraboo was as foreign as she seemed. The incongruence between her European features and her exotic language and behaviour were a particular source of puzzlement. So too was her sudden appearance in the village. But as the search for answers continued, and as well-travelled and knowledgable guests were invited to interrogate the woman, there was, finally, a breakthrough. A Portuguese man called Manuel Eynesso came to meet Caraboo. He had spent time in

----
105

the Far East, and as a consequence was able to identify her speech as a mixture of Sumatran dialects. Though he was not fluent in the language, Eynesso gave a basic translation of her story. She was, he said, a woman of some importance – a princess, perhaps – who had been kidnapped from her home on the island of Javasu in the East Indies, then transported to England against her will. Furnished with this information, another man was found who had more intimate knowledge of the region from which she had come. He conversed with Caraboo on several occasions, then provided the Worrals with a detailed description of her background.

According to this account, Caraboo's mother was from Malaysia and her father, though white in complexion, was from China. He was a powerful man, to whom the people of Javasu would kneel in deference. Of the island itself she explained that the waters around it were very shallow, so that large vessels were unable to approach. This, presumably, was the reason it was thus far unknown in the West. Of the goods produced there, she named cassia, rice and white pepper. The seas surrounding Javasu, she said, were home to flying fish.

Thereafter, the Worrals' guest was made especially welcome at Knole Park. She was, after

all, an Oriental princess. And with that role now established, Caraboo delighted her hosts and their numerous visitors with strange behaviour. She made bows and arrows, and could fence with great skill; she wore feathers in her hair and played a tambourine; she danced outlandishly and swam naked. Her fame spread. A portrait was painted, showing her in a white turban and flowing, golden gown. Newspapers published the story, together with the picture. And that was her undoing.

Princess Caraboo is often described as a hoaxer and a fraud, but that is not entirely fair. Though she was recognised in the newspaper by a former landlady, and thereafter exposed as Mary Willcocks, from Witheridge in Devon, the identity she had assumed was not of her own making; it was bestowed upon her. Mary was a poor and restless woman. She had been forced to give up a child the previous year, and the child had then died. She had spent time travelling with Romani travellers, from whom she learned some of her vocabulary. And though clearly intelligent, she was also undoubtedly disturbed. Her refusal to speak English was, in part, a kind of withdrawal from the world. She did not seek fame, nor did she cheat the Worrals out of money or possessions. She came to them against her will and tried more than once to escape.

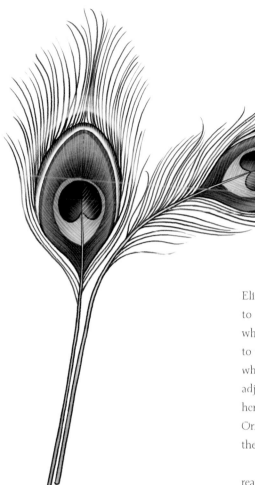

Elizabeth Worral longed for an exotic solution to the mystery of Caraboo, and that's precisely what they gave her. Mary, in turn, just played up to the role. She listened in to the words of those who thought she could not understand, and she adjusted her behaviour accordingly. She found herself in the midst of a society obsessed with the Oriental, and she reflected that obsession back at them. She became a kind of mirror

Following the embarrassing revelation of her real identity, Mary was sent to the United States in the company of three chaperones chosen by Mrs Worral. She was greeted there as a celebrity. Seven years later she returned to England and tried, for a time, to live off her now fading fame. But it didn't last. She began to wander again, in Spain and in France, before marrying and finally settling in Bristol. She died there, aged 75, in 1864, and is buried in the city in an unmarked grave. Javasu was never heard of again.

Mary's story, in virtually every detail, was provided by others – by Manuel Eynesso and by the second man (who was unnamed in contemporary accounts). Both men most likely assumed that Mary was indeed a foreigner, and that she would therefore be unable to comprehend or contradict their inventions.

O N PAGE FIVE OF the *Literary Gazette*, published in London on 12th February 1825, a brief notice gave details of a discovery in the South Pacific. Tucked between a paragraph on the use of Chinese costumes in French theatre and the announcement of an invention that allowed people to breathe while in dense smoke, were the following words:

> New Island—*Captain Hunter, of the merchant-vessel Donna Carmelita, is stated, in the New South Wales' Journals, to have discovered a new island in the Southern Ocean, in July last. The latitude is 15 degrees 31'S. and longitude 176 degrees 11'E.*

The notice went on to explain that the crew of the **Donna Carmelita** 'had friendly intercourse with the King and natives', who, it stated, 'do not seem to differ from the South Sea Islanders, already known to navigators'.

As it happened, the *Literary Gazette* was out of date. The discovery of Onaseuse, as it was called, had in fact been made two years previously, on 20th July 1823. The journal's brief outline of the story also failed even to hint at the extraordinary level of detail about the island and its inhabitants with which Captain Hunter had furnished the world's media.

Phantom islands are rarely explored, for obvious reasons. But this one was. And the information recorded by captain and crew demonstrated beyond any doubt that Onaseuse was not a case of mistaken identity. There were no other known islands in this region (around 250km northwest of Fiji), and certainly none that matched the descriptions given by those onboard the *Donna Carmelita*.

In the *Asiatic Journal* of May 1824, the chief officer's record of the discovery was published, together with Captain Hunter's. It explained how, in the early afternoon, the crew had landed on the island, which was 'fruitful and very populous', and been met there by a group of natives. Given that they had never seen Europeans before, the locals were remarkably welcoming. One came quite willingly onboard the ship, bringing them 'refreshments', including meat and plantains.

Some of the men then went ashore and were taken to meet the king of the island, who was called 'Funafooah'. We went without weapons, wrote the chief officer, 'as it would make them have more confidence in us'. He went on:

> *Most of them were armed with war clubs, with short round heads, some with spears from 24 to 40 feet long, afterwards I saw some much longer. A great number of women, many of whom carried two spears, as I judged for the use of the men.*

Despite this abundance of weaponry, there was no sign of aggression from either

Onaseuse

side. (Though one cannot help but wonder how useful a 40-foot spear would be anyway, other than to decorate a tall tale.)

The relaxed conviviality of this first contact went on, according to the report, with the king and chief officer exchanging a variety of gifts with one another. Funafooah and his brother were presented first with white shirts, which they wore, seemingly delighted. In return, they offered 'a hog, a basket of yams and bananahs and cocoa-nuts'.

Next, 'I made him a present of a looking glass, which seemed to surprise them greatly; it went from the King to the Queen, and from her all round, every one taking a look at it, and then touching the crown of their heads with it, that ceremony they performed with every little thing given them. He took a shell from his neck and gave it me.'

While this extended ritual of gift-giving was fascinating enough, the elements of the story that aroused most interest among anthropologists were those describing the islanders' physical features. There were, for instance, the red, circular tattoos that many of the women had inscribed on their arms, and the fact that most wore shells as ornaments, and had some kind of white paste smeared on top of their hair.

But of particular interest was the observation that 'all the women and men had their little fingers cut off at the second joint on the left hand, and the women had their cheek bones perforated, and the blood smeared round about an inch; I suppose the mark of beauty.' This kind of self-mutilation had never been recorded anywhere else, so it was not surprising that others were keen to visit the island again, to learn more about these unusual people.

The problem was, nobody could find it. Numerous attempts were made to relocate Onaseuse throughout the nineteenth century, but it was never seen again. Which begs the obvious question: was it ever there in the first place?

According to the chief officer, the island 'was entirely composed of lava, in some places, almost a metal', which leaves open the possibility that it may have been destroyed in a volcanic explosion, some time after that initial visit. If so, it would not be the first such disappearance in the South Pacific. But there is something not quite right about this explanation. The detail seems rather too convenient and deliberate, as though the chief officer knew, as he wrote it, that such an excuse would be necessary. As though he knew that Onaseuse was going to disappear.

The story of this island, with its warlike-yet-friendly natives, and their familiar-yet-extraordinary appearance, is suspicious. This region was well travelled by the early nineteenth century, and no other report of an island in this area was ever recorded. It is within the bounds of possibility that Onaseuse – or Hunter Island, as the captain renamed it – did exist. But more likely is that the captain and his crew, eager to make names for themselves, invented every last detail.

ROBERT PEARY DID NOT always tell the truth. That much is certain. For most of the twentieth century he was believed to be the first man to reach the North Pole, in April **1909**, with his companion Matthew Henson and four Inuit assistants. After a very public and very nasty campaign, his rival Dr Frederick Cook's claim to have got there one year earlier was dismissed, and Peary took the

Crock

title. But doubts over his account continued to niggle, and in the **1980**s the British explorer Wally Herbert was asked to settle the matter. He was given access to diaries and records from the expedition, and spent three years examining the evidence. Herbert concluded that, though he may have been close, Peary never made it to the Pole.

Whether or not the American knew he had failed is impossible to say, but if he did lie it would not have been the first time. He had form. On a previous journey in the north, Peary claimed to have sighted land beyond Axel Heiberg Island, at around 83 degrees north, which he called Crocker Land. It was a canny choice of name. George Crocker was one of Peary's financial backers, from whom he wished to squeeze

that the new island did not exist, those friends jumped to his defence. This was to be a lie with tragic consequences.

The University of Illinois, together with the American Geographical Society and the American Museum of Natural History, organised an expedition to explore Crocker Land. Its leader, Donald Baxter MacMillan, was filled with confidence: 'Its boundaries and extent

er Land

some cash to fund his next expedition. Flattery, he thought, would do the trick.

Crocker, though, did not pay up, and had it not been for Peary's frantic race to the Pole with Frederick Cook, that might have been the end of the matter. Crocker Land would have been forgotten. But Peary had influential friends, and when Cook claimed in 1909

can only be guessed at,' he said, 'but I am certain that strange animals will be found there, and I hope to discover a new race of men.'

The expedition did not discover a new race of men, but MacMillan did, briefly, think he had found Peary's island. At the end of April 1914, nearly a year after setting out, he sighted land in the Arctic Ocean. He could see hills, valleys and snow-capped mountains up ahead. This, he thought, was surely the place. His guide was not convinced. It was 'mist', the Inuit said: a mirage or *fata morgana*. MacMillan scoffed and the men carried on, the ice around them breaking up as summer advanced. It was a dangerous journey, and a foolish one, for the Inuit was right. Five days and **125** miles later, MacMillan was forced to concede that Crocker Land was not there. 'We were convinced that we were in pursuit of a will-o'-the-wisp,' he wrote, 'ever receding, ever changing, ever beckoning . . . My dreams of the last four years were merely dreams; my hopes had ended in bitter disappointment.'

The return journey was to prove no more successful. Having stopped in northern Greenland to carry out scientific research, one of the Americans, Fitzhugh Green, murdered the Inuit Piugaattoq – a crime for which he was never charged, and showed little remorse. Among the other members of the expedition, Walter Ekblaw was struck with excruciating snow blindness and Maurice Tanquary had both big toes amputated after frostbite. All came close to starvation. And though two relief missions were sent to the Arctic to rescue them, both failed. It was not until August **1917** – more than four years after setting out – that the last of the men finally got home.

For Peary, the findings of the expedition might have seemed like good news. The mirage seen by MacMillan provided a convenient excuse for his non-existent island. He too, he could claim, had been fooled by 'mist'. But unfortunately for the explorer's reputation, while he had lied to the world, he had not lied to his diary. In his book, *Nearest the Pole*, Peary had given the dates at which he sighted Crocker Land. But his journal, examined later, contradicted the published account. There is no mention of the island anywhere. So either he remembered what he had seen only later, or he invented it. And though he remains a hero to many in the United States, who still allow him the benefit of the doubt, the former explanation hardly seems likely.

So while Frederick Cook had lost his title as the first to reach the North Pole, he had succeeded at least in catching out his rival. But that small achievement was itself undermined, for in his account of that journey, on which he crossed the sea ice north of Axel Heiberg Island, Cook had gone further. Crocker Land did not exist, he wrote, but there was another island out there, which he called Bradley Land. A detailed account of this discovery was offered, including two photographs as proof. Like Peary, Cook had named his island after a sponsor, and like Peary he had made it up. The photographs were fake. The race for the Pole had been desperate, and neither man came out of it well. It was not until **1969** that anyone reached the top of the world by foot. And that person was Wally Herbert.

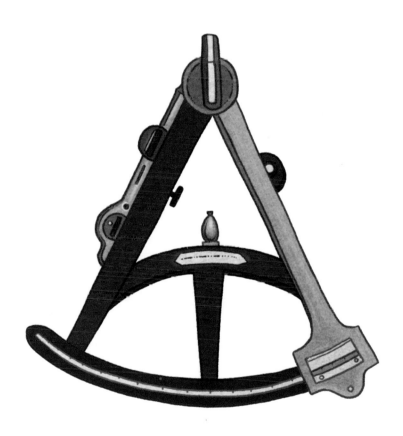

Recent Un-Discoveries

Recent Un-Discoveries

IN 1875, THE BRITISH Royal Navy decided it was time for a tidy up. They knew their charts of the Pacific were littered with inaccuracies, and Captain Sir Frederick Evans was tasked with putting them right. In total, Evans deleted **123** phantom islands from the Admiralty's maps (though three of these later turned out to be real). It was a significant achievement, and a sign of just how many errors had been lingering unnoticed. But it was far from the end of the story.

If the preceding centuries had been an age of great geographical discoveries, the twentieth was largely a time of un-discovery, when virtually all the remaining ex-isles were finally expunged. Many of these, understandably, were in the Arctic and Antarctic. These were most difficult regions in which to travel, and the last to be properly explored. They were also the places where optical illusions such as *fata morgana* were most liable to confuse weary sailors, and where enormous icebergs were sometimes hard to distinguish from tiny islands.

For a long time there was good reason to leave uncertain islands, shoals and reefs on the map, even after doubts had been raised. Such things could be a real danger to shipping, and it was better to be cautious than to be sorry. But when navigational technology finally made it possible to determine a location precisely, this began to change. And in the latter half of the twentieth century, when satellites revolutionised our view of the world, one could finally check an island's location without the inconvenience of actually having to visit.

Today the era of new island discoveries is over, and the age of un-discovery is likewise coming to an end. But that convenience is accompanied by loss. For millennia our oceans have been populated by imagined islands, reflecting back at us something about our understanding of the world. But now these places are endangered and headed for extinction. We are paying for our cartographic completeness with a feeling that something, somewhere, is missing.

L OS JARDINES SHOULD not have survived for as long as they did. As phantom islands go, they are among the most inexplicably stubborn. In the four hundred years or more in which they remained on the map, the islands changed size, shifted their location by twelve degrees of latitude, and shrank from ten to just two. They could never have been all that they were supposed to be, and in the end they were nothing at all.

But perhaps it was that very ability to transform themselves that saved the islands for so long. When they couldn't be found in one place, they moved to another; and when they still could not be seen, they became smaller. For century after century, mariners and cartographers gave them the benefit of the doubt. So while other Pacific phantoms were

erased one by one, Los Jardines stood firm, the tiny letters E.D. – existence doubtful – sometimes appended to their name like a badge of honour. It was not until the Second World War that they began to disappear from charts, and not until 1973 that the International Hydrographic Bureau finally let go of them altogether. They had had a long and restless life.

The islands were first mentioned by Álvaro de Saavedra, the cousin of Hernán Cortés, destroyer of the Aztec Empire. Saavedra was employed by Cortés to undertake an expedition from New Spain to the Indonesian

Los Jardines

Though he died in 1529, Saavedra left behind an account of that journey, describing his exploration of the coast of New Guinea and his discovery of several new islands in the western Pacific. Among these finds was an atoll or a group of low lying islands that he named Los Buenos Jardines (the good gardens) on account of their fecundity.

The archipelago was located to the northeast of New Guinea, at somewhere between eight and twelve degrees north. On arrival, Saavedra and his crew were welcomed by the heavily tattooed inhabitants. The Spaniards were greeted with music and singing, and were led to a house to meet the chief. The islanders gave them presents, including 2,000 coconuts, and the crew remained as guests for eight days while their ailing captain rested.

About fifteen years after this visit,

Maluku Islands in 1527. Despite losing two of the three vessels that set out on that voyage, Saavedra succeeded, and in doing so became the first European sailor to cross the Pacific Ocean from east to west.

another Spaniard, Ruy Lopez de Villalobos, also came upon a group of islands – ten, in this case – that he called Los Buenos Jardines. Whether these were the same islands as those found by Saavedra is not quite clear, for this part of the Pacific is well populated by islets, atolls and other specks of land, and in those days there was little certainty when it came to establishing one's location. It is entirely possible that these Jardines may have been somewhere else altogether, but were given the same name for the same reason.

Though firmly established on the map for the next two hundred years, Los Jardines could easily have been forgotten. Countless islands and archipelagos were 'discovered' several times over and given multiple names, only one of which could ultimately survive, and by the eighteenth century it was becoming clear that whatever Los Buenos Jardines had once been, the name had loosened its grip on the land. The next logical step was for it to be erased.

But what happened next was not logical; it was almost completely inexplicable. For rather than disappear altogether, as ought to have happened, two pieces of extraordinarily creative cartography gave Los Jardines a new lease on life, and allowed them to remain undisturbed for another two centuries.

The first of these occurred when, for reasons not at all clear, the islands migrated northwards to between 20 and 23 degrees of latitude, and became two rather than ten. This happened at some point in the mid-eighteenth century, and was probably the result of mistaken identity, though by whom and for what is unknown. For several decades thereafter the islands were in limbo, appearing sometimes in the north, sometimes in the south, sometimes in neither position, and occasionally both. But by the end of that century, their fate was certain: they had left the south behind.

At this point it should have taken only a few failed attempts to find Los Jardines before the error was realised. But instead, one massive error was compounded by another, when an over-imaginative mapmaker was given the logs of two British vessels, the *Scarborough* and the *Charlotte*. These ships were part of the fleet that took the first British convicts to Australia, and were captained by John Marshall and Thomas Gilbert, both extremely good and respected navigators. In 1788, the two left Australia heading for China, and en route discovered several islands and atolls, some of which now bear their names. But according to their logs, the *Scarborough* and *Charlotte* made a rather strange about-turn while in the region of 22 degrees north and 150 degrees west. Presumably, with crews suffering from scurvy, they were making a quick, rather desperate search for Los Jardines. But neither captain claimed to have found them.

Somehow, though, that over-imaginative mapmaker thought otherwise. While plotting the course of the two ships and noting the peculiar manoeuvre at this location, it was assumed that the missing islands had been found. Marshall was credited with the discovery, and lingering doubts over the existence of Los Jardines were suddenly (and wrongly) abandoned.

Over the next **150** years, some mariners claimed to have seen the two islands, while others failed to find them, and gradually the doubts returned. But the reputations of Marshall and Gilbert were such that Los Jardines stayed put. They survived the thorough culling of phantoms in **1875**, and they even survived, on some maps at least, the Second World War, when this region of the Pacific was particularly busy.

The irony, of course, is that Los Buenos Jardines *did* exist. They were, most likely, one of the atolls – Enewitok or Bikini – that lie closest to the original location given by Saavedra. Quite why they took on a life of their own is not entirely clear. But at some point Los Jardines turned from real islands into phantom islands, and then finally into ex-isles.

Terra
Nova
Islands

IN THE HISTORY OF polar exploration, it is often those who have failed most spectacularly who have been lionised. Sir John Franklin is among the most famous of British explorers, though he didn't find the Northwest Passage, and he and 128 of his men died (and probably ate each other). Captain Scott and Ernest Shackleton are national heroes, though neither man achieved exactly what they set out to do. Perhaps this is why the name of Phillip Law is not better known. For his is not a story of heroic failure, it is one of almost unmitigated success.

Throughout the twentieth century, Australia was at the forefront of efforts to explore and map Antarctica, and the legacy of that work is clear. Today, the country claims around 42 per cent of the entire continent – a territory that, at more than two million square miles, is only twenty per cent smaller than Australia itself. Phillip Law is in no small part

----

responsible for that legacy. Appointed as director of the Australian National Antarctic Research Expeditions (ANARE) in 1949, Law established the first two of the country's permanent research stations – Mawson and Davis – and negotiated the transfer of the third, Wilkes, from the United States, thereby ensuring an Australian presence on the continent that continues to this day.

He led 23 expeditions in his career, and succeeded in mapping more than 3,000 miles of coastline and almost 400,000 square miles of the interior. He visited parts of the continent no person had ever seen before, and did so with a level of courage and determination that few could match. His was an astonishing record, and he was rewarded with a CBE, a Polar Medal, the Founder's Gold Medal of the Royal Geographical Society, and numerous other honours. Law knew that what he had done was something quite special. 'Scott, Shackleton, Mawson and such men,' he once boasted, 'I explored ten times as much as all of them put together.'

After leading ANARE until 1966, Law retired to more sedate roles back home, though he continued to serve as chairman of the Australian National Committee on Antarctic Research for many years. Looking back on his time in the far south, he was often known to describe himself as 'one of the last people in the world who's had the joy of new exploration.'

He died, aged 97, in 2010.

In a career as distinguished as this one, some small mistakes are both inevitable and, surely, forgivable. In Law's case those mistakes came in the form of the Terra Nova Islands, which he discovered on 8th March 1961. These two scraps of land were spotted by the crew of the *Magga Dan*, 14 nautical miles north of Williamson Head, on Oates Coast, East Antarctica.

In his record of that expedition, Law made little of the discovery, noting only that: 'I was able to take some interesting bearings of features along the coast and also discovered two small islands about 8 miles to the west of the ship's position. I called these the Terra Nova Islands after Pennell's ship' (the vessel that had brought Captain Scott south on his final, fateful journey).

The islands' existence was accepted in April 1970 by the United States Board on Geographical Names, and a brief note in the board's archives sets out their position, together with a few details of the discovery.

It was not until almost 20 years later that anyone went looking for the Terra Nova Islands again. After all, there was no particular reason to seek out two rocks in the half-frozen sea. But in February 1989, a German scientific expedition, GANOVEX V, was working along Oates Coast and took the opportunity to visit

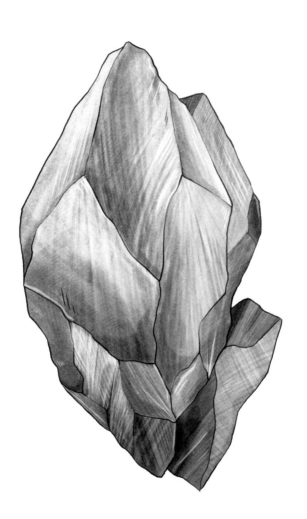

these unexplored islands. Their geologists were helicoptered out to the location to map them and to take rock samples. What they found, or didn't find, surprised them.

In a telex sent from their ship, the *Polar Queen*, shortly afterwards, the fruitless search for the Terra Nova Islands was described as an 'interesting discovery that shows how incompletely known parts of the Antarctic coast still are today, or how much less secure the "known" is'. The telex was sent by Dr Norbert W. Roland, a scientist on board, who explained that they had good reason for assuming the islands would be there. They were noted, after all, in the *Antarctic Pilot*, used by nearly all shipping in the area. The telex went on:

> *As the islands could not be located in previous days, either during helicopter flights along the coast or from on board the GANOVEX V ship* Polar Queen, *which had approached on the evening of* 22.2.89 *to two miles off the position of the islands, I made an expedition flight in the morning of the 23rd*

February together with the captain of the Polar Queen, *Peter Brandal, and the pilot Trevor McGowan. An area of more than* 15 x 20 km² *was searched from* 5000ft *in quadrants. The unlimited view permitted us to see a much larger area, and whereas other islands like the Aviator Islands that lie in the east and are according to the maps much smaller, or Babushkin Island that lies in the south, were clearly seen and recognised, and even the ship could be clearly seen from more than* 10km *distance from the pack ice, there was no trace of the Terra Nova islands.*

The *Polar Queen* later made depth measurements with an echo sounder, confirming ocean depths of 170–355 metres in the supposed location of the islands. Dr Roland suggested that Law had perhaps confused two icebergs for islands. 'In summary,' he wrote, 'the islands don't exist and have to be crossed out of the official maps.'

IN 1997, THE GOVERNMENTS of Mexico and the United States began to negotiate a treaty on the countries' maritime limits in the Gulf of Mexico. It was a return to a tricky issue, which had been discussed in the late 1970s but never fully settled. There were, specifically, two parts of the Gulf that did not come within the 200 nautical mile limit of either country, and so remained in dispute. They were known as Hoyos de Dona, or the Doughnut Holes.

The negotiations over these areas might have been simpler and less protracted had it not been for one important detail: the Doughnut Holes were thought to contain significant reserves of oil and gas, so control over them could be worth a lot of money indeed. Naturally, then, work was required to establish exactly where the land owned by each country extended to, and therefore how the areas might best be divided. From 1997, the two governments concentrated on the westernmost of the Doughnut Holes (which covers nearly seven thousand square miles) since the eastern region was also bordered by Cuban waters, making it considerably more complicated.

According to older charts of the Gulf, the closest piece of Mexican land to the disputed zone was a small island called Bermeja, around 100 miles off the coast of the Yucatan Peninsula. The problem was that when a vessel from the Mexican navy was sent to find it, the island was nowhere to be seen. The country's claim had to be shifted, consequently, some distance to the south.

Ber

As it turned out, there was little controversy over this disappearance at the time. At least in public. The negotiations continued and, in the year 2000, were finalised. Concerns that drilling close to the border could lead to one side profiting from the other's oil or gas reserves, however, led to a ten-year moratorium on extraction being agreed. It was not until 2009, as the end of the moratorium approached, that the question of Bermeja was raised again. And this time people wanted answers.

Bermeja had appeared on maps for a long time. Its first mention came in the early sixteenth century, with details given not only of its location and size, but also its appearance (the name Bermeja is derived from a Spanish *and Gulf of Mexico*, published by the United States Hydrographic Office in **1885**, Bermeja is only briefly mentioned. And for good reason. 'This island is found on all the old charts', the book explains, 'but its existence is more than doubtful. The officers of the Spanish navy, in their search for the *Negrillo*, saw nothing of it, neither did another officer who purposely

meja

word used to describe the island's reddish colour). It continued to be shown on maps for several hundred years, always in the same place and conforming to roughly the same shape. Up until the nineteenth century, no doubts were ever raised about its existence. But by the end of that century things had changed.

In *The Navigation of the Caribbean Sea* looked for it in **1804**. Captain E. Barnett, R.N., also closely examined this neighbor-hood in **1844**, without meeting with anything.' The rest of the world, it seemed, had long considered Bermeja to be lost. Only its owners had failed to notice.

By **2009**, however, Mexican politicians and journalists were certainly taking notice.

And they were incredulous. How could a piece of the country's territory have simply vanished? Opposition leaders in the senate demanded to know what had happened. How could the island have been allowed to disappear? And how could the former president, Ernesto Zedillo, have been so complacent? He had ceded territory unnecessarily to the United States, they claimed, and in doing so had potentially handed billions of barrels of oil to their northern neighbour.

As the accusations flew, conspiracy theories began to circulate. Many Mexicans suggested that, far from a cartographical error or natural disaster, the disappearance of Bermeja had been a deliberate act – the work of the CIA, on behalf of its government. America had stolen Mexican oil, they said, by destroying the island with a bomb.

Journalists too smelled something fishy in the story. Looking back at the original negotiations with the US in the late 1990s, they found that papers relating to these discussions were missing. The minutes of debates on the treaty were no longer available, and no one could explain where they were. The suggestion was raised that perhaps President Zedillo himself had been involved in the disappearance of Bermeja, in a secret, corrupt deal with American oil firms.

Such theories were given more fuel when it was recalled that only one senator had opposed Zedillo's stance at the time. José Angel Conchello had argued that Mexico should claim the disputed Hoyos de Dona for itself, and had gone on to condemn the exploratory work being done by US companies in the region. This was Mexico's water and Mexico's oil, he had said. But Conchello was killed in an unexplained car accident while negotiations were still ongoing. Eleven years later there were calls, from all sides, for an investigation.

Answers to the question of Bermeja were clearly needed, and to try and find them two more comprehensive surveys were commissioned in 2009. These surveys were intended to discover whether the island had sunk, and if so how. They would also explore the wider area in the Gulf, to see if it might just have been misplaced.

Those surveys were completed in the spring of that year, with researchers from several universities involved in searches by sea and by air. Many hundreds of square miles were covered, and soundings of the ocean floor were taken. But the scientists found nothing at all. Bermeja wasn't there, and most likely never had been.

No further investigations have taken place, either into the disappearance of the island or the events surrounding the treaty with the United States in the year 2000.

IN NOVEMBER 2012, the *Southern Surveyor*, a research vessel from Australia, was in the Coral Sea west of New Caledonia. The scientists on board were studying the tectonic evolution of the region, but took a break from their work to investigate a rather peculiar anomaly. They had noticed that an island indicated on some of their maps was not present on the nautical chart they were using. According to the chart, the ocean was never less than **1,400** metres deep in that area, yet the maps – and Google Earth – indicated otherwise. A roughly oval stretch of land, **15** miles long by three miles wide, was clearly shown, alongside its name: Sandy Island.

The researchers approached the stated position with some caution. After all, they had no idea exactly what to expect. A half-submerged sandbar, a reef or shoal: such hazards are no less real in the twenty-first century than they ever were before, and the ambiguity of the available information meant that hidden dangers were a real possibility. But in the end their caution proved unnecessary. The ocean floor remained stubbornly in place, more than a kilometre beneath them. The ship sailed right through the middle of Sandy Island, and Sandy Island wasn't there.

Within a few days, the world's media were relating the details of this un-discovery to their readers. The *Sydney Morning Herald* gleefully announced 'The mystery of the missing island', while the *Guardian* called it 'The Pacific island that never was'. Online, it became one of the most widely-shared news stories of the year. But why? The non-existence of a small, uninhabited piece of land in a remote corner of the world is hardly significant, politically or geographically. Yet somehow the idea of a place that was on the map and yet was not real caught the public's imagination.

In the weeks following its un-discovery, scientists and journalists scrambled for answers. How could an island have slipped through the net of satellite technology for so long? Where did this ex-isle originate? At first it was assumed that a computer error was to blame: a blip of pixels somewhere in the system. But no, there it was on a British Admiralty map of **1908**, in the same place and the same shape as it still was **100** years later.

Sandy Island was first reported by a whaling ship in **1876**. It was a single sighting, and most likely a simple error. It was later marked on some maps, though not others, and was occasionally flagged up as a potential phantom.

Sandy Island

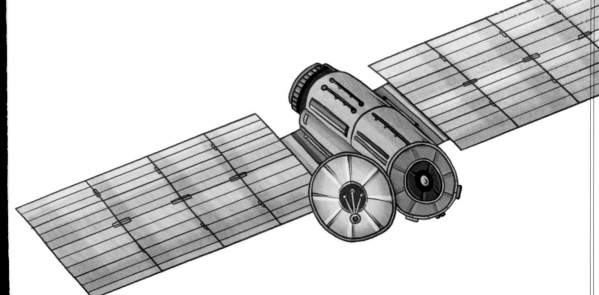

But when the US military digitised their charts, the island found its way into the World Vector Shoreline Database, which is widely used by scientists and in mapping software. In this rarely visited part of the ocean, that database is still far from perfect.

Had it been there, Sandy Island would have belonged to France, located as it was within the territorial waters of New Caledonia, a 'special collectivity' of the country. But the French were not much concerned about the disappearance; they had removed it from their charts more than three decades earlier. As had the Americans. In fact, as it turned out, Sandy Island had been un-discovered more than once already.

Most recently, in the year 2000, a group of amateur radio enthusiasts had noticed its absence while travelling to the Chesterfield Islands, 100km to the west. But while their testimony had received little attention at the time, in 2012 the response was swift. By the end of November, the National Geographic Society had declared that the island would no longer appear on their maps, and other publishers soon followed suit.

But Sandy Island has not disappeared altogether. Look carefully on Google Earth between Australia and New Caledonia, where the island was once thought to lie. At first there is nothing to be seen, but zoom in a little and the island's yellow outline reappears: an empty shape on the ocean's surface.

For a time, fantastical photographs appeared inside that shape, added by the programme's users. Green valleys, harbours, waterfalls, forests, a beach resort, even an atomic explosion: the space that Sandy Island once inhabited was filled with the imaginings of people the world over. The photographs were meant to be funny, of course, but they also express a kind of delight that such a space could still exist in the world. Now, however, those photographs too are disappearing.

For millennia, explorers had edges to reach and to go beyond. They had blanks to fill and *terra incognita* to discover. Always, they had mysteries to solve. Part of the joy of geography was the knowledge that there was still more out there, just waiting to be found. Today, with maps on our computers and our phones, and with satellites circling above us, it seems all that has gone forever. The science of navigation has worked towards the eradication of uncertainty and the end of mystery, and to an astonishing degree it has succeeded. We can know where we are and what direction we are travelling with just the click of a button. And though that technology brings its own kind of wonder, part of us mourns what has been lost.

The story of Sandy Island offered people hope. It showed that mystery had not been destroyed entirely; it is still out there, if you know where to look. There may be no more unknown islands to be found in the world, but perhaps there is another ex-isle still intact, a phantom waiting to be un-discovered. And perhaps we should leave it that way.

Other Un-Discovered Islands

Hundreds of islands have come and gone over the centuries, in our stories and on our charts. This book has introduced only a small selection of them. Gathered below are ten additional islands, just waiting to be explored.

Buyan This Slavic myth has clear echoes of classical and Celtic stories. The island is a place of happiness and eternal life, which can appear and then disappear again. Some versions of the tale describe Buyan as the source of all weather, where the winds have their home. It has been linked to the real island of Rügen – now part of Germany – in the Baltic Sea.

Mayda Like Hy Brasil, the island of Mayda appeared early on European charts; and like Hy Brasil, there is no clear explanation of why it is there. Most often crescent-shaped, it was first marked in the Atlantic west of Brittany, but gradually drifted towards North America. Despite never being visited or claimed by any country, Mayda appeared on a Rand McNally map as late as 1906.

Isle of Demons A relatively short-lived phantom, the Île des Démons (often shown as a pair of islands) was mapped off the coast of Labrador or northeastern Newfoundland from the early sixteenth into the mid-seventeenth century. The name was probably supposed to refer to Quirpon Island and Belle Isle, off the northern tip of Newfoundland, which were widely believed to be haunted.

Elizabeth Island Discovered by one of the greatest British explorers, Francis Drake, in 1578, Elizabeth Island lay at the southern tip of South America. It was one of the very first overseas claims by Britain (along with Martin Frobisher's in Baffin Island the previous year). But unfortunately, nobody was quite certain what piece of land the name actually referred to, so the claim meant little.

Rupes Nigra Gerardus Mercator's map of the northern polar regions, published posthumously in 1595, is a beautiful and ambitious piece of cartography. At its centre is Rupes Nigra, a mountain as tall as the clouds, surrounded by an enormous ocean whirlpool. As far as speculative geography goes, a magnetic mountain at the North Pole at least had a degree of logic on its side. But, like much of that map, it had no basis in reality.

Saxemberg In the middle of the South Atlantic, about 600 miles northwest of Tristan da Cunha, Saxemberg was first reported by a Dutch sailor in 1670. After disappearing for more than a hundred years, it was seen again – though in a slightly different position – at least twice in the early nineteenth century. Then it disappeared forever. Whether these reports were errors or lies, it is impossible to be sure.

Dougherty Island One of several non-existent islands in the far south of the Pacific Ocean, Dougherty was seen at least three times between 1841 and 1886. But that was the end of it. Captain Scott looked for it in 1904, but found nothing, and John Davis made a thorough and unsuccessful search in 1909, when he also un-discovered the Nimrod Islands, Emerald Island and the Royal Company's Islands.

Podesta First sighted by the marvellously named Captain Pinocchio in 1879, in the Pacific Ocean far to the west of Chile. The island was small, the captain reported, with a circumference of less than a mile, and rose to just forty feet above sea level. It also didn't exist, and was gradually erased from the charts throughout the twentieth century.

Kantia Plato is not the only philosopher to be linked with a phantom. When the trader Johann Otto Polter found a new island in the Caribbean in 1884 he named it after Immanuel Kant and claimed it for Germany. Unfortunately, Kantia was never seen again. Despite searching for it numerous times without success over the next 25 years, Polter refused to accept he had been mistaken.

Maria Theresa Reef There are a handful of phantom reefs in the South Pacific still waiting to be fully expunged from the maps. Hydrographers shifted this one more than 1000km eastwards in September 1983, but that didn't make it any easier to find. It does not appear on Google Earth, but is still included on some paper charts.

References

The Isles of the Blessed Homer, trans. Samuel Butler, *The Odyssey*, New York, 1898. Plato, trans, B. Jowett, 'Gorgias', in *The Dialogues of Plato*, Oxford, 1871. Geoffrey of Monmouth, trans. Basil Clarke, *Vita Merlini*, University of Wales Press, 1973.

Kibu A. C. Haddon, W. H. R. Rivers, C. G. Seligmann, A. Wilkin, *Reports of the Cambridge Anthropological Expedition to Torres Straits: Volume 5*, Cambridge, 1904.

Hawaiki Te Ahukaram Charles Royal, 'Hawaiki: The significance of Hawaiki', Te Ara: the Encyclopedia of New Zealand, www.TeAra.govt.nz/en/hawaiki/page-1.

Hufaidh Gavin Maxwell, *A Reed Shaken by the Wind*, Longmans, Green and Co, 1957. Wilfred Thesiger, *The Marsh Arabs*, Longmans, Green and Co, 1964

Thule Strabo, trans. Horace Leonard Jones, *Geography*, Vol. I of the Loeb Classical Library edition, Harvard University Press, 1917. Polybius, trans. Evelyn Shuckburgh, *The Histories of Polybius*, London, 1889.

St Brendan's Island Trans. Denis O'Donoghue, *Brendaniana: St Brendan the Voyager in Story and Legend*, Dublin, 1893.

Frisland John Dee, manuscript in the British Library, quoted in *John Dee: The World of an Elizabethan Magus* by Peter J. French, Routledge, 1972.

Davis Land Lionel Wafer, *A New Voyage and Description of the Isthmus of America*, London, 1704.

The Auroras Edgar Allan Poe, *The Narrative of Arthur Gordon Pym of Nantucket*, New York, *1838*. Log of the *Atrevida*, quoted in James Weddell, *A Voyage towards the South Pole*, London, *1825*.

Atlantis Plato, trans. H. D. P. Lee, *Timaeus and Critia*s, Penguin, 1971. Ignatius L. Donnelly, *The Antediluvian World*, New York, 1882.

Island of Buss George Best, *A True Discourse of the Late Voyages of Discoverie*, London, 1578. Thomas Wiars, quoted in *The Principall Navigations, Voiages, Traffiques, and Discoueries of the English Nation* by Richard Hakluyt, London, 1589.

Sarah Ann Island Asenath Taber, 'Diary, December 3, 1854 – September 2, 1855', The Mariners' Museum Library Gallery, librarygallery.marinersmuseum.org/items/show/43.

Lemuria or Kumari Kandam Philip Sclater, 'Some Difficulties in Zoological Distribution' in *The Nineteenth Century 4*, 1878. Philip Sclater, 'The Mammals of Madagascar' in *The Quarterly Journal of Science*, April 1864.

Crocker Land Donald Baxter MacMillan, *Four Years in the White North*, London, 1918.

Terra Nova Islands Phillip Law, quoted in the *Independent*, 16 May 2010. Phillip Law, quoted in *The Scotsman*, 11 March 2010. Phillip Law, quoted in *Antarktis* by Norbert Roland, Spektrum Akademischer Verlag, 2009. Telex sent by Dr Roland from the *Polar Queen*, quoted in *Antarktis* by Norbert Roland, Spektrum Akademischer Verlag, 2009. (Translated from the German by Anja Hedrich.)

Further Reading

-------- ---

General William H. Babcock, *Legendary Islands of the Atlantic*, American Geographical Society, **1922**.

Donald S. Johnson, *The Phantom Islands of the Atlantic*, Souvenir Press, **1997**.

Raymond H. Ramsay, *No Longer on the Map*, Ballantine Books, **1973**.

Henry Stommel, *Lost Islands*, University of British Columbia Press, **1984**.

Individual Islands Hundreds of books have been written about Atlantis, and dozens more about Lemuria. The vast majority of these can probably be read as fiction.

Joanna Kavenna's *The Ice Museum* (Penguin, **2006**) is an excellent introduction to Thule, both as a place and as an idea. Barry Cunliffe's *The Extraordinary Voyage of Pytheas the Greek* (Walker & Company, **2002**) provides a more scholarly approach to the subject.

Andrea di Robilant's *Venetian Navigators* (Faber & Faber, **2011**) offers a very readable account of Frisland and the other Zeno islands, though the author seems rather too eager to believe the tale.

Barbara Freitag's *Hy Brasil: The Metamorphosis of an Island* (Rodopi, **2013**) peels away the many falsehoods and misconceptions that surround this island.

About the Author

Malachy Tallack is an author and singer-songwriter who has written for many publications. He is contributing editor of the online magazine, *The Island Review*, and his previous book was *Sixty Degrees North*. Tallack is from Shetland, Scotland, and he lives in Glasgow.

About the Illustrator

Katie Scott is the illustrator of *Botanicum* and of the bestselling *Animalium*, which was chosen as the *Sunday Times* Children's Book of the Year, 2014. She studied illustration at the University of Brighton and is inspired by the elaborate paintings of Ernst Haeckel.